THE LAST YEARS OF
SOUTH WEST STEAM

PETER TUFFREY

GREAT NORTHERN

ACKNOWLEDGEMENTS

I am grateful for the assistance received from the following people: Roger Arnold, David Burrill, John Chalcraft, Paul Chancellor, David Christie, Peter Crangle, John Law, Hugh Parkin, Bill Reed, Andrew Warnes, Sue Warnes.

Gratitude should also be expressed to my son Tristram Tuffrey whose support has been invaluable throughout the course of the project.

Great Northern Books
PO Box 1380, Bradford, BD5 5FB
www.greatnorthernbooks.co.uk

© Peter Tuffrey 2023

Every effort has been made to acknowledge correctly and contact the copyright holders of material in this book. Great Northern Books Ltd apologises for any unintentional errors or omissions, which should be notified to the publisher.

All rights reserved. No part of this book may be reproduced in any form or by any means without permission in writing from the publisher, except by a reviewer who may quote brief passages in a review.

ISBN: 978-1-914227-54-7

Design and layout: David Burrill

CIP Data
A catalogue for this book is available from the British Library

INTRODUCTION

South West England is the largest of the nine Regions of England, consisting of over 9,000 square miles. Yet, the area is third from last in terms of population which hampered the development of a number of railway schemes in the second half of the 19th century. Nevertheless, a rich network was established featuring several duplicate lines and branches that were later swept away by Dr Beeching.

The counties contained in the South West Region are: Cornwall; Devon; Dorset; Gloucestershire; Somerset; Wiltshire. Some of the towns and cities present include: Bath; Bristol; Bournemouth; Cheltenham; Cirencester; Dawlish; Exeter; Exmouth; Gloucester; Liskeard; Newton Abbot; Penzance; Plymouth; Salisbury; Sidmouth; Swindon; Tavistock; Truro; Wadebridge; Weymouth; Yeovil.

As with other areas of Britain, the initial schemes had industrial interests pushing the projects forward. Perhaps the earliest to adopt steam was the Bodmin & Wadebridge Railway which opened in 1834. This transported mainly china clay from Wenfordbridge to Wadebridge on the coast. There was also a passenger service offered from Bodmin to Wadebridge nearly from the outset, though this later became infrequent. The route subsequently formed part of the London & South Western Railway's system.

The first main line arrived in the area from London to Bristol as the Great Western Railway was completed in 1841. At the time, Bristol was one of the most important harbours in the country and businessmen there were concerned about losing this role to other places, such as Liverpool, which benefitted from a rail connection. These people were also interested in joining Bristol and Exeter. Just before the opening of the GWR route, the first section of the Bristol & Exeter Railway between Bristol and Bridgwater saw trains running. The B&ER had entered an agreement with the GWR for the latter to run the line for an annual rent and small levy on goods and passengers. Three more stages were necessary before completion was reached with the opening of Exeter St Davids station on 1st May 1844. At this time, the two lines (GWR and B&ER) had a combined mileage approaching 200 which was the longest in the country.

The GWR did not build the Paddington-Bristol line without competition. The London & Southampton Railway hoped to branch from Basingstoke to Bristol but lost out to the GWR. The L&SR was completed in May 1840 and the company soon expanded the area served. A branch to Gosport was laid in 1842 and another connection was made via ferry with Portsmouth. Local rivalries forced the L&SR to make a name change to the London & South Western Railway.

As the GWR and associated companies forged westward, the L&SWR was keen to compete. A project to join Southampton with Poole and Dorchester came to Parliament in 1845 and was ready for traffic just two years later. Though originally independent, the company was soon absorbed by the L&SWR. At Dorchester, an agreement with the Wilts, Somerset & Weymouth Railway, which was subsequently taken over by the GWR due to financial problems, allowed the L&SWR passage to Weymouth. This section was both broad and standard gauge.

Around this time, the L&SWR was also involved in a link to Salisbury. The company was not too far away as the route approached Southampton and a branch was built from Bishopstoke, opening in 1847. The arrival of the L&SWR in Salisbury was thought to be the precursor for an immediate extension to Yeovil and Exeter. Yet, opposition from the GWR and challenging financial conditions delayed the plans. These were later revived in the mid-1850s with a direct line from Basingstoke to Salisbury which cut out the longer route via Bishopstoke. The Salisbury & Yeovil Railway was also formed independently but with backing from the L&SWR. As two thirds of the route to Exeter was planned, the L&SWR soon had work in hand on the final section from Yeovil to Exeter. The line from Basingstoke reached Salisbury in 1859, then Yeovil in June 1860 and trains ran throughout to Exeter in July 1860. A new station was built at Queen Street, though two years later a connection was made with St Davids station.

As the railway reached Exeter from Bristol in the early 1840s, the South Devon Railway was actively promoting a route from there to Plymouth via Newton Abbot. Authorised in 1844, the company had subscribers including the GWR, Bristol & Exeter Railway and Bristol & Gloucester Railway. Isambard Kingdom Brunel was appointed engineer but he held misgivings on the suitability of contemporary engines for the difficult terrain of the area. Brunel decided to use an atmospheric railway where locomotives used a partial vacuum to operate. This offered advantages of cost savings and greater traction. The line to Newton Abbot was ready in mid-1846 though the atmospheric apparatus was not installed for several months. In the meantime, locomotives were rented from the GWR. When the system was ready, successful operation was achieved for a time before higher than anticipated costs surfaced and unreliability during bad weather occurred. By the end of 1848 the atmospheric railway was abandoned. Construction of the remaining route occurred in two sections: Totnes in July 1847; Plymouth during April 1848.

The completion of the London & Southampton Railway hurt Falmouth which was historically the port of arrival for mail from around the world bound for London. The L&SR offered a quicker route and the service soon switched ports. In the early 1840s, the Cornwall Railway wanted to build a straight route from Falmouth to London without stopping at some of the important towns and cities. This was soon proved impractical and the promoters turned to connecting with Plymouth. Delayed until the mid-1850s, the Plymouth-Truro line was ready at the end of the decade and Falmouth in 1861.

A mineral line – the Hayle Railway – was built in the late 1830s to the west of Truro and in the 1840s plans were formulated to use this as a connecting point for a route from Penzance to Truro. The West Cornwall Railway was ready for traffic in 1852. When the Cornwall Railway made a connection in 1860 there was a clash of line gauges as the WCR adopted standard and the CR had broad gauge. In the mid-1860s, the WCR was obliged to lay broad gauge lines but was not financially able to do so, leading to a takeover by the GWR and associated companies. Ten years later, the GWR amalgamated several interests to become the dominant force in the area.

The L&SWR was not content to settle in Exeter and in the early 1850s took steps to take over lines from Exeter to Barnstaple and Bideford. The lines were acquired in the mid-1860s and a branch was built to Plymouth in 1874. This provided a departure point for routes to Holsworthy (1879) and Bude (1899) as well as connecting with the long-held Bodmin & Wadebridge Railway. The company was then well-positioned to compete in key areas of South West England.

Another major line in the eastern half of the area was the Somerset & Dorset Railway which had started as two independent companies but came together in the early 1860s. By the middle of the decade, the route stretched from Burnham-on-Sea to Bournemouth via Templecombe. Towards the 1870s a plan to connect with Bath via the Somerset coalfield was formulated and was completed in just a few short years. The financial cost of this led to the Midland Railway and L&SWR to bail out the S&DR to form the Somerset & Dorset Joint Railway.

With the main lines built, connections to places missed followed. These included branches to: Swanage; Lyme Regis; Seaton; Sidmouth; Exmouth; Brixham; Kingswear; Newquay; Looe; St Ives; etc.

At Grouping in 1923, the lines were either operated by the GWR or Southern Railway, whilst at Nationalisation, these fell into the Western or Southern Region of British Railways respectively. In the 1950s and early 1960s, many of the routes quietly enjoyed their final years before the end of steam, the Beeching cuts and dieselisation. South West England was well-served by the standard classes of the GWR, including: Collett's 'Castle', 'King', 'Hall', 'Grange' and 'Manor' Class 4-6-0s; the numerous 5700 and 8750 Class 0-6-0PTs; 5101 and 6100 2-6-2Ts; Churchward 2800 Class 2-8-0s; Churchward 4300 2-6-0s. Also, the locomotives of the L&SWR and Southern Railway were employed: Bulleid 'West Country' Pacifics; Maunsell's 'N' Class 2-6-0; Drummond T9 Class 4-4-0; Adams 415 Class 4-4-2T; Drummond M7 Class 0-4-4T. Mainly on the S&DJR, BR's Standard Classes were employed: Class 5 4-6-0; Class 4 4-6-0; Class 4 2-6-0; Class 9F 2-10-0.

South West England was particularly affected by the closures in the 1960s, owing to the rural nature of the area and decline of passenger and freight usage. Many of the branches were lost, along with the S&DJR and part of the L&SWR west of Exeter. Yet, some of the branches have found a new lease of life as heritage railways populated by former local engines owing to the strong representation of the GWR in preservation. Some of the heritage railways are: Bodmin & Wenford; West Somerset; Dart Valley; Swanage; Helston; Somerset & Dorset Railway Heritage Trust; Tarka Valley line. Hopefully, these can continue to draw support and build new networks to keep steam in South West England for years to come.

Peter Tuffrey
Doncaster, June 2023

Above ASHCHURCH STATION – NO. 42400
A local train is headed by Fowler 4P Class 2-6-4T no. 42400 at Ashchurch station on 29th August 1960. Just nine days earlier, the engine arrived for work at Saltley shed. Photograph by Bill Reed.

Below ASHCHURCH STATION – NO. 47539
Another Fowler locomotive has a short local service at Ashchurch station. No. 47539 of the 3F Class is seen on 29th August 1960. Photograph by Bill Reed.

Above **AXMINSTER STATION – NO. 30584**
The London & South Western Railway expanded westward from Yeovil to provide competition for the Great Western Railway. The line ran to Exeter and was opened in 1860. Around halfway, the route skirted the western side of the town of Axminster and a station was erected for the opening. The inhabitants of the coastal town of Lyme Regis were keen to be served by the railway but adverse terrain and the associated cost were factors against the L&SWR being induced to take the project. Help arrived in the form of the Light Railways Act 1896 as this reduced the capital necessary to build lines, particularly for rural areas. In 1899, the Axminster & Lyme Regis Railway was formed with around £80,000 of funds for the route which took four years to build. The connection was made with the L&SWR at Axminster and the company had agreed to operate the line whilst also providing some of the capital investment. The six-mile route was used for just over sixty years. In the branch bay at Axminster station is Adams 415 Class 4-4-2T no. 30584 during the late 1950s. Photograph by Bill Reed.

Opposite above **AXMINSTER STATION – NO. 73082**
British Railways identified the need for a mixed traffic locomotive in the mould of Stanier's Class 5 4-6-0 for use across the country. The Standard Class 5MT 4-6-0 appeared between 1951 and 1957 with 172 engines built. No. 73082 was the product of Derby in June 1955 and delivered to Stewarts Lane depot, London. Some ten years have elapsed from this event to the image of the engine at Axminster station with an express in November 1965. During the late 1950s, 20 Standard Class 5s employed by the Southern Region were named, including no. 73082 which became *Camelot*. These were taken from withdrawn Urie N15 'King Arthur' Class 4-6-0s. Following withdrawal, no. 73082 was rescued from Woodham Brothers' scrapyard and was restored for use on the Bluebell Railway. Photograph courtesy Rail-Online.

Opposite below **AXMINSTER STATION – NO. 30582**
William Adams designed the 415 Class 4-4-2T for suburban services around London. All were built by contractors between 1882 and 1885 when the total was 72. By the turn of the century, new classes had been introduced and the 415s were displaced to other parts of the system. This coincided with the opening of the Lyme Regis branch and two class members were sent there. In the 1930s, the Southern Railway was to scrap the engines working the route, yet a suitable replacement could not be found and no. 30582 was one of the pair kept on the line. A third 415 Class engine was found to augment the duo in the mid-1940s and these remained to 1961. No. 30582, which is at Axminster station, was almost preserved but went to the scrapyard. Photograph by Bill Reed.

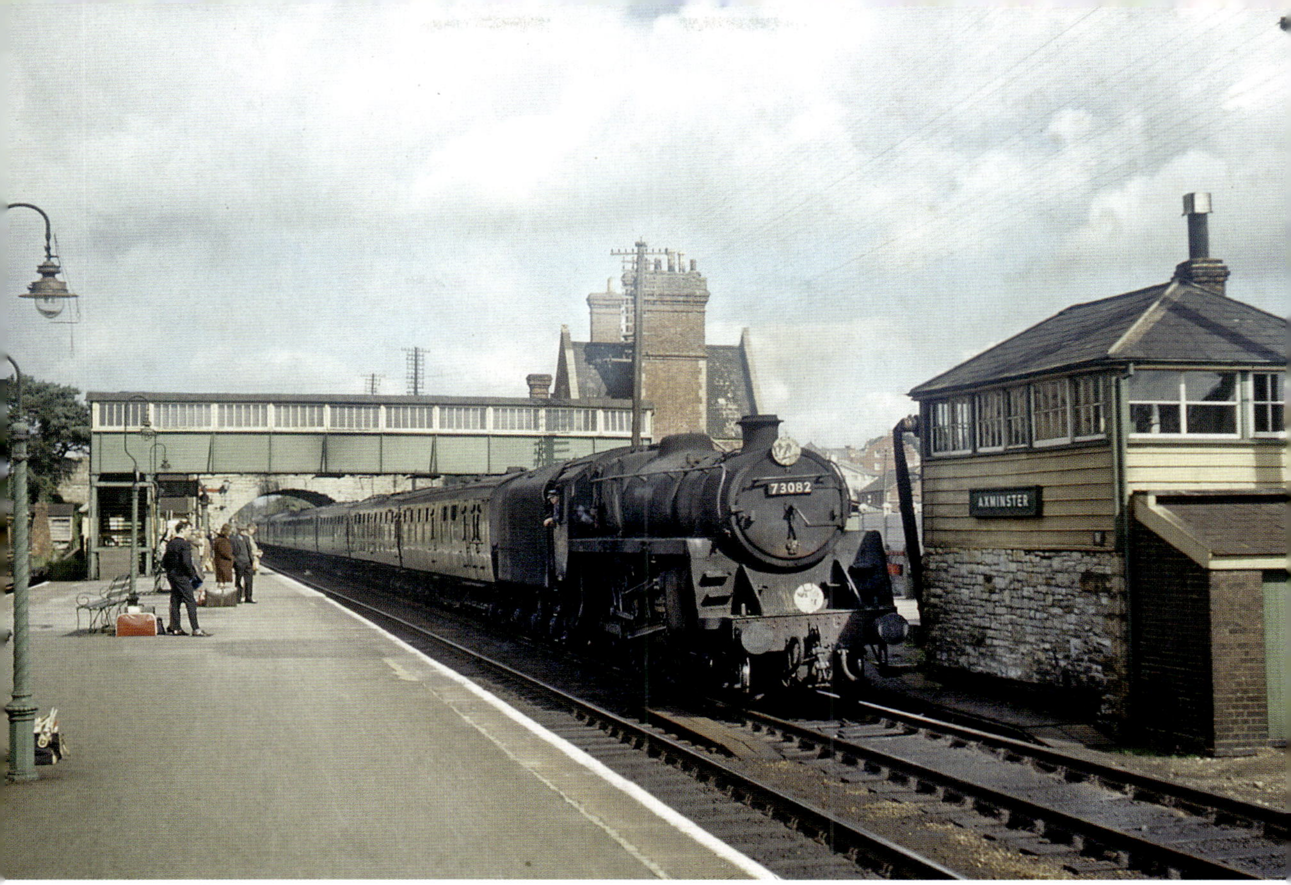
73082
AXMINSTER

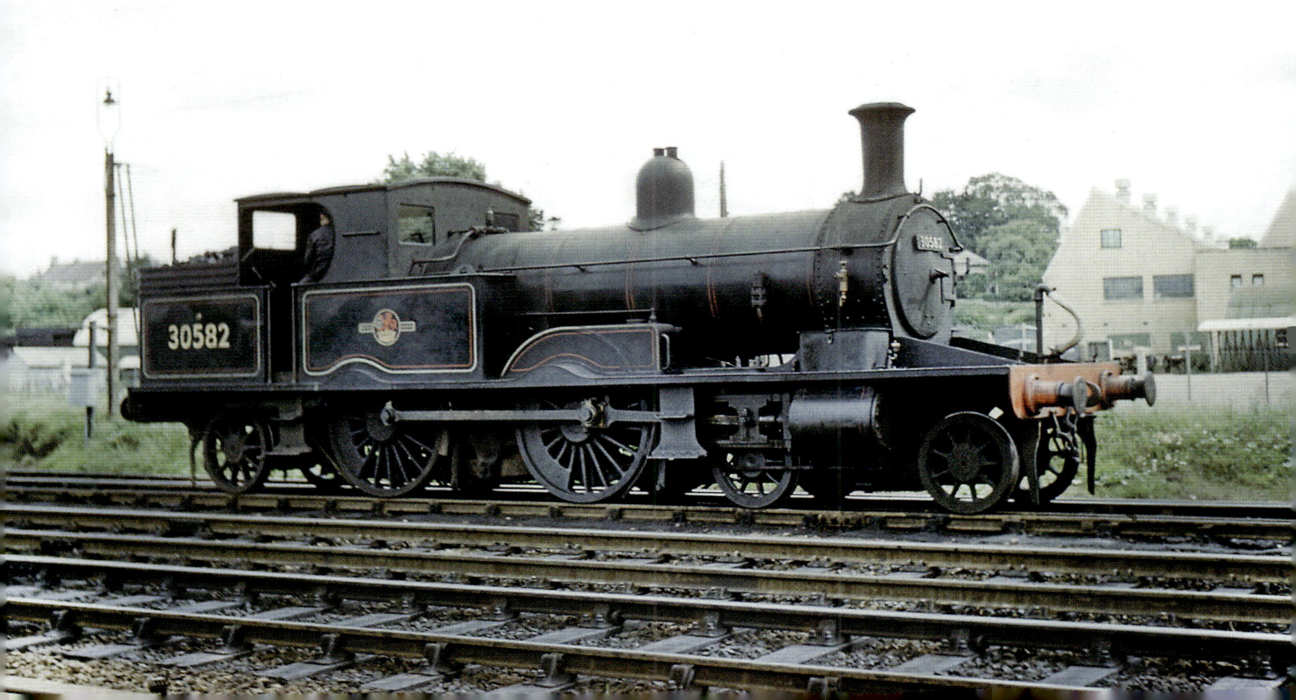
30582
30582

Above **BARNSTAPLE – NO. 34080**
The North Devon Railway arrived at Barnstaple in 1854 as part of the project to connect Exeter and Bideford. Just over ten years later, the L&SWR took over the company and another decade passed before the branch to Ilfracombe was ready for traffic. This had to cross the River Taw and a bridge was built to connect the original station, which became Barnstaple Junction, and the new one Barnstaple Quay, later Barnstaple Town. Bulleid 'Battle of Britain' Class Pacific no. 34080 *74 Squadron* is crossing from north to south in 1963. The Ilfracombe branch closed in 1970 and the bridge was removed later in the decade. Photograph by P.C. Wakefield courtesy Colour-Rail.

Opposite above **BARNSTAPLE STATION – NO. 30254**
Dugald Drummond designed the M7 Class 0-4-4T for suburban traffic, mainly around London, at the end of the 19th century. These were built at the L&SWR's Nine Elms and Eastleigh workshops, though the greater share was with the first mentioned. Of the 105-class total, 95 locomotives were built there, with no. 30254 amongst the number, being an early example appearing in August 1897. The construction period lasted to 1911. The locomotive has stopped with a train at Barnstaple station. For much of the 1950s, no. 30254 was locally allocated. Photograph by Bill Reed.

Opposite below **BARNSTAPLE STATION – NO. 41314**
A modern tank engine was necessary at the end of the Second World War for use on the London Midland & Scottish Railway. H.G. Ivatt produced the 2MT 2-6-2T in 1946 and these locomotives were erected into the 1950s. Under BR, the design was thought useful for other regions and 30 of the last engines built were allocated for use across the Southern Region. No. 41314 was new to Exmouth Junction in May 1952 and moved between several depots in the mid-1950s before settling at Barnstaple for six years. The locomotive is at the station c. 1960. Photograph by Bill Reed.

Above BATH GREEN PARK STATION – NO. 73012
Of the 172 Standard Class 5 locomotives built, 130 came from Derby and 42 from Doncaster. No. 73012 was erected at the aforementioned in August 1951. Leeds Holbeck received the engine at this time though two years later it was sent to the Western Region. No. 73012 is with a local train at Bath Green Park on 9th September 1963 when allocated to Swindon. Photograph courtesy Colour-Rail.

Opposite above BATH GREEN PARK STATION
From the mid-1840s, the Midland Railway had a connection between the industrial Midlands and the South West through the merger with two companies that had created a line from Birmingham to Bristol. The important city of Bath was just a short distance away and the MR sought to build a line there in the mid-1860s. Leaving the original line at Mangotsfield, the new route ran for 10 miles to a terminus next to Green Park. The station, originally known as Bath, was designed by the company's architect John Holloway Sanders, who was sympathetic to the surrounding Georgian style. The Somerset & Dorset Joint Railway also used the facility from 1874. British Railways changed the name to Bath Green Park and this is displayed prominently on the frontage here on 21st July 1962. The station was open to passengers for just another four years though freight traffic lasted to the early 1970s. Since that time, the buildings have found use as a place for local businesses. A small fire damaged the Grade II-listed train shed in April 2023. Photograph by B.W.L. Brooksbank.

Opposite below BATH GREEN PARK STATION – NO. 73049
An engineman takes a break before the 15.25 Bristol to Bournemouth train leaves Bath Green Park station on 5th September 1964. At the head of the formation is BR Standard Class 5 4-6-0 no. 73049. The last engine of the second batch built, consisting of 20, no. 73049 was originally sent to work at Leicester Midland shed when new in December 1953. Around 18 months later, a transfer to Bath Green Park occurred and the locomotive joined several other class members for service on the S&DJR, in addition to the Birmingham route. After five years in Bath, no. 73049 moved on to Shrewsbury and Bristol but returned to Bath between July 1962 and October 1964. The last depot for the locomotive was Oxford and withdrawal occurred in March 1965. Photograph by L. Rowe courtesy Colour-Rail.

Above BATH SPA STATION
The Great Western Railway reached Bath in 1840 as the main line from London moved towards Bristol. Bath Spa was in the eastern half of the city on the bank of the River Avon, whilst Green Park was further west. The entrance of Bath Spa is bustling here on 21st July 1962. Photograph by B.W.L. Brooksbank.

Below BATHAMPTON STATION – NO. 6850
Fresh from works attention at Swindon, Collett 'Grange' Class 4-6-0 no. 6850 *Cleeve Grange* has a local train to Bristol passing a waiting track gang at Bathampton station on 19th August 1958. Photograph by Hugh Ballantyne courtesy Rail Photoprints.

Above **BERE ALSTON STATION – NO. 41316**
A local train is at Bere Alston station – on the L&SWR branch to Plymouth – with Ivatt Class 2MT 2-6-2T no. 41316. The engine was based in Plymouth from 1956 to 1963. Photograph by Bill Reed.

Below **BERE ALSTON STATION – NO. 41295**
A mineral line existed to the west of Bere Alston and with the opening of the branch thoughts turned to creating a better connection. To keep costs down, a Light Railway Order was granted for a short line to Callington which opened in 1908. Ivatt Class 2MT 2-6-2T no. 41295 has a Bere Alston-Callington train in the early 1960s. Photograph by Bill Reed.

Above BERE ALSTON STATION – NO. 75025
A smaller version of the Standard Class 5 4-6-0 was necessary to serve a wider area and the Class 4 was developed. No. 75025 was one of 80 class members. The engine has a southbound freight at Bere Alston on 18th April 1964. Photograph by B.W.L. Brooksbank.

Below BERE ALSTON STATION – NO. 30034
Drummond M7 Class 0-4-4T no. 30034 approaches Bere Alston on 11th August 1961. Plymouth-based at this time, the locomotive was there for another year before moving on to Nine Elms, London. Photograph by Bill Reed.

Above **BERE ALSTON STATION – NO. 30225**
The predecessor to Drummond's M7 Class 0-4-4T was William Adams's O2 Class 0-4-4T which was built by the L&SWR from 1889 to 1896. No. 30225 was constructed at Nine Elms Works in November 1892. The engine is with a Callington branch train at Bere Alston, c. 1960. Photograph by Bill Reed.

Below **BERKELEY ROAD STATION – NO. 1426**
Just south of Gloucester on the line to Bristol, Berkeley Road station was the location for a branch to Sharpness Dock in 1876. At the end of the decade, the Severn Railway Bridge opened to connect with Lydney and South Wales. Collett 1400 Class 0-4-2T no. 1426 has a Lydney branch train at Berkeley Road. Photograph by Bill Reed.

Above BERKELEY ROAD STATION – NO. 73164

The Bristol & Gloucester Railway opened Berkeley Road station as Dursley & Berkeley with the line in July 1844. The name was changed in the following year and lasted to the station's closure on 4th January 1965. Berkeley Road was between two junctions. The northern one led to Dursley, as a two-mile branch from the main line was built in 1856, whilst the second one to Sharpness was opened in 1876. BR Standard Class 5 no. 73164 approaches Berkeley Road with the 17.20 Gloucester to Bristol local on 18th July 1964. Note on the right, Lydney has been removed from the station nameboard. The Severn Railway Bridge was damaged by a vessel colliding with a pier and though initially to be repaired, a second collision resulted in a demolition order for the bridge. The connection was severed not to be reopened. Photograph courtesy Colour-Rail.

Opposite above BINEGAR STATION – NO. 75072

The Somerset Central and Dorset Central Railways joined forces in 1862 to create the Somerset & Dorset Railway. By the end of the decade, the company had decided to expand northward to connect to Bath and Bristol. This line ran from Evercreech via Midsomer Norton to Bath where the company obtained powers to use the Midland Railway's branch. A number of engineering challenges existed on the route which exhausted the S&DR's financial resources. The company soon had to be rescued from collapse and was taken over by the MR and L&SWR as a joint enterprise. Binegar station opened with the extension in July 1874 with modest facilities to serve the local village but was important for banking engines to be provided for tackling Masbury summit. A local train is at Binegar's southbound platform behind BR Standard Class 4 no. 75072 on 2nd September 1961. Photograph from the Dave Cobbe Collection courtesy Rail Photoprints.

Opposite below BODMIN ROAD STATION – NO. 4565

The Cornwall Railway, which ran between Plymouth and Falmouth, reached completion in mid-1859. A branch to Bodmin was originally planned but opposition from a local landowner and costs stopped this project and Bodmin Road station was made the permanent stop for the town several miles away. Not until the late 1880s did the GWR finally build the branch which also connected with the ex-Bodmin & Wadebridge Railway then operated by the L&SWR. Churchward 4500 Class 2-6-2T no. 4565 has a local train at Bodmin Road. Under BR, the locomotive worked at St Blazey depot and was condemned there in October 1961. Photograph by Bill Reed.

Above BODMIN GENERAL STATION – NO. 5502

The branch from Bodmin Road to Bodmin General station opened in 1887, with the latter as a terminus. In the following year, a connection was made to the L&SWR which provided a difficulty as through trains then had to reverse. Bodmin General had a goods shed and cattle dock, as well as small engine shed. Both passenger and freight services lasted to 1967. The station later became the hub for the Bodmin & Wenford heritage line. Collett 4575 Class no. 5502 has a passenger train at Bodmin General in the mid- to late 1950s. Photograph by Bill Reed.

Opposite BODMIN GENERAL STATION – NO. 4552

G.J. Churchward developed an earlier 2-6-2T design to achieve slightly better speeds on their mixed traffic duties in the early 20th century. The 4500 Class first appeared in 1906 from Wolverhampton Works. Production continued to just after Grouping when 75 had been constructed. No. 4552 was amongst a batch built early in the First World War, being towards the end of the group of 15. The locomotive was in traffic for March 1915 and the working life spanned 46 years. Goods are being loaded on to a train behind the engine at Bodmin General station on 1st September 1960. Photograph by Bill Reed.

Above **BOURNEMOUTH CENTRAL STATION – NO. 34028**
Rebuilt Bulleid 'West Country' Pacific no. 34028 *Eddystone* has a passenger train at Bournemouth Central station c. 1960. Photograph by Bill Reed.

Below **BOURNEMOUTH SHED – NO. 30863**
A pair of Maunsell 'Lord Nelson' Class 4-6-0s is in the yard at Bournemouth shed. Featured is no. 30863 *Lord Rodney* whilst on the right is no. 30850 *Lord Nelson*. Photograph by Bill Reed.

Above **BOURNEMOUTH SHED – NO. 30782**
Over a servicing pit at Bournemouth shed is Urie N15 'King Arthur' Class no. 30782 *Sir Brian*. The locomotive was allocated there 11 years, from 1951 to 1962 when condemned. Photograph by Bill Reed.

Below **BOURNEMOUTH SHED – NO. 34102**
Bulleid's 'West Country/Battle of Britain' design had several flaws which saw a number of class members rebuilt. No. 34102 *Lapford* was the last of the group in traffic that was not rebuilt. The engine is at Bournemouth shed, c. 1960. Photograph by Bill Reed.

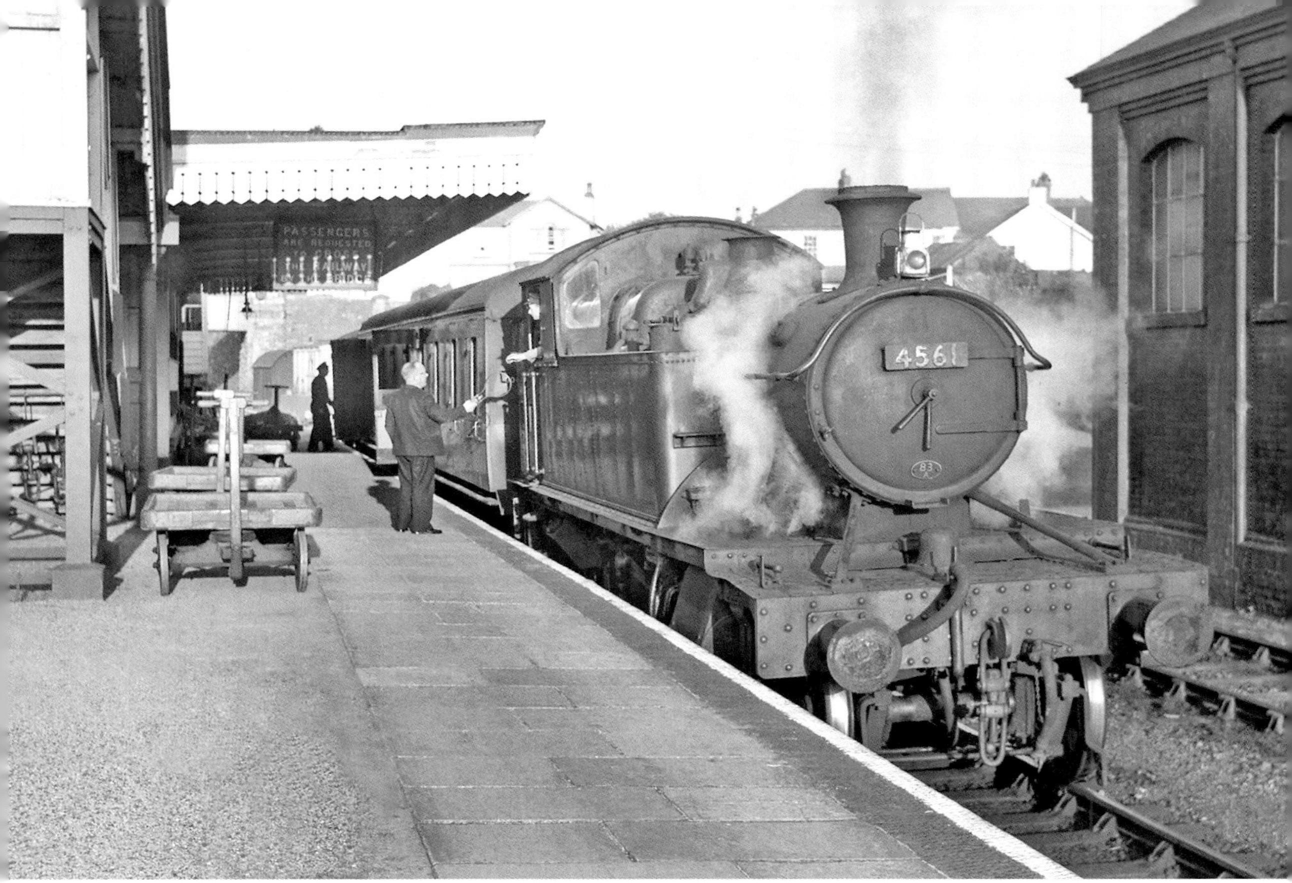
PASSENGERS
ARE REQUESTED
4561

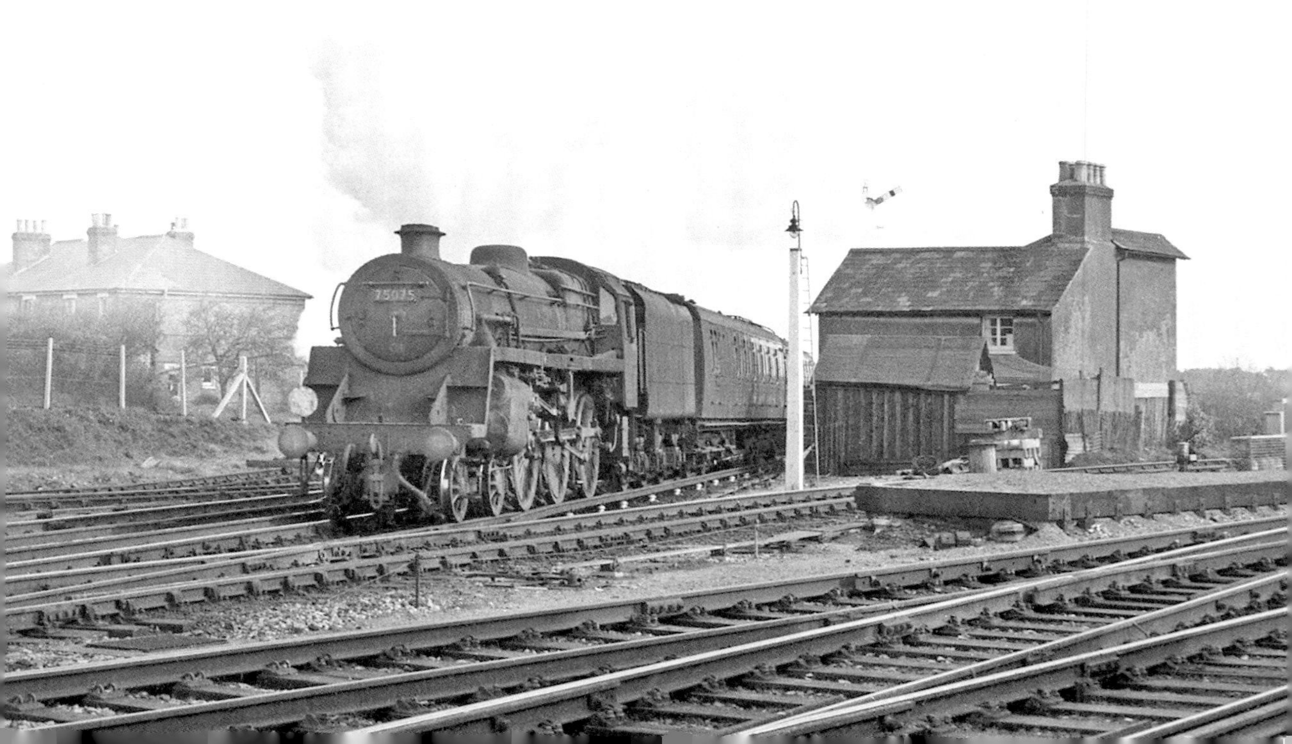
75075

Above BRADFORD-ON-AVON STATION – NO. 7924

An ambitious project to link Wiltshire, Somerset and Dorsetshire by a main line, along with several branches, was promoted in the early 1840s. Construction of the Wilts, Somerset & Weymouth Railway began later in the decade and the first section opened in 1848. Financial difficulties were then encountered and the company was sold to the GWR in 1850. A branch to Bradford-on-Avon was partially completed before this but did not open until 1857 and only occurred owing to legal action started by interested parties. Hawksworth 'Modified Hall' no. 7924 *Thorneycroft Hall* has stopped at Bradford-on-Avon with the 13.00 Salisbury to Bristol local on 10th August 1963. Photograph by Hugh Ballantyne courtesy Rail Photoprints.

Opposite above BRENT STATION – NO. 4561

The South Devon Railway was promoted as early as the 1820s, yet the project did not reach parliament until 1840. Starting at Exeter, the route hugged the coastline to Newton Abbot, then turning inland to reach Plymouth. Just west of Totnes on the line was Brent station, which served the village of South Brent from mid-1848. The SDR missed several places and branches to reach these were built subsequently. In the 1880s, the GWR took over a company founded to connect Kingsbridge with the main line at Brent. Covering 12 miles, this first saw traffic in 1893. Passenger trains were operated in the main, latterly holidaymakers in the summer months, with some freight and a goods shed was present at Brent. The latter is partially visible on the right – interestingly this still survives in use as a dentist surgery. Churchward 4500 Class 2-6-2T no. 4561 has the 19.30 train from Kingsbridge on 2nd June 1961. The branch was in use to 1963, whilst Brent station closed in 1964. Photograph by Tony Cooke courtesy Colour-Rail.

Opposite below BRANKSOME – NO. 75075

A relatively late arrival of the railways occurred at Bournemouth. The place was just a small coastal village before the Southampton & Dorchester Railway was built to the north in 1847. Then, Bournemouth was promoted as a holiday resort and spa. The first branch arrived from Ringwood in 1870 followed by a second branch from Poole in 1872, which also later accommodated the Somerset & Dorset Railway. A connection between the two was established at Branksome in the 1880s. BR Standard Class 4 no. 75075 is running off the junction for Bournemouth Central station and is heading westwards towards Branksome station. The lines to Bournemouth West station are in the bottom right. Of note is the house right in the corner of the junction which apparently still stands and incredibly has not been affected by the passing traffic. Despite the closure of Bournemouth West, the junction still exists owing to a depot being located on the line. Photograph by S.P. Sutton courtesy Colour-Rail.

Above **BRENT STATION – NO. 4561**
No. 4561 was one of the post-Grouping Churchward 4500 Class members. Built at Swindon in October 1924, the engine was in service over 37 years, as withdrawal occurred during May 1962. This event was around 18 months away on 2nd September 1960 when the locomotive was pictured at Brent station. At the time, no. 4561 was allocated to Newton Abbot, though in the following July transferred to Plymouth Laira which took the engine to the end. Photograph by Bill Reed.

Opposite above **BRIDGWATER STATION**
A view of Bridgwater station on 22nd April 1963 during a lull in activity. Built by the Bristol & Exeter Railway in 1841, the station was originally the terminus for the route as the undertaking was completed in stages. The full route was operational from 1844 under the GWR, with a formal takeover occurring at the end of the decade. Most of the buildings have survived, yet sadly the Railway Hotel has been demolished. Photograph by B.W.L. Brooksbank.

Opposite below **BOURNEMOUTH WEST STATION**
As mentioned, lines to Bournemouth were developed in the 1870s. A branch from Poole to Bournemouth was completed early in the decade and used by the Somerset & Dorset Railway to reach Bournemouth. Used from 1874, Bournemouth West station closed in September 1965 as a temporary measure but permanently from the following month. The frontage has been captured on 20th April 1963. Photograph by B.W.L. Brooksbank.

Opposite above BRIDPORT STATION – NO. 5507
Engineman and signalman prepare to exchange the line staff at Bridport station in April 1958. The promotion of the Wilts, Somerset & Weymouth Railway inspired people in Bridport to make a connection with the line which passed to the east. In the mid-1850s an Act was obtained to build nine miles of line, making a junction at Maiden Newton and this was achieved in late 1857. The GWR leased the route from opening to the early 20th century when taken over completely. Though marked for closure by Dr Beeching, local opposition delayed this for over ten years to 1975. Collett 4575 Class no. 5507 is the locomotive. Photograph by Bill Reed.

Opposite below BRIMSCOMBE BRIDGE HALT
With the introduction of steam railmotors in the early 20th century, several halts were opened to serve local passengers. One of these was Brimscombe Bridge Halt which was located on the Swindon, Cirencester, Gloucester and Cheltenham line. Opened on 1st February 1904, the halt was in use to 1964. The route had been in existence from the 1840s as the Cheltenham & Great Western Union Railway promoted the project but was only able to fulfil a portion from Swindon to Cirencester. The remaining part was built by the GWR. Photograph by B.W.L. Brooksbank.

Below BRIDPORT STATION – NO. 7782
Charles Collett's 5700 Class 0-6-0PT was the GWR's most numerous design with 863 examples erected from 1929 to 1950. Yet, over the years several sub-classes were created as variations were introduced. The 5700 Class designation covered 300 locomotives built over just two years in a relatively unusual move by the GWR to employ contractors to build their locomotives. No. 7782 was amongst 25 ordered from Armstrong Whitworth & Co. Ltd and appeared from late 1930 to early 1931. The engine is at Bridport station in April 1958. Photograph by Bill Reed.

THE LAST YEARS OF SOUTH WEST STEAM

Above **BRIMSCOMBE SHED – NO. 6106**
Brimscombe was close to Sapperton incline and was the location for banking locomotives to wait for their services. Collett 6100 Class no. 6106 is perhaps on call here at the shed in September 1964. Photograph by David Christie.

Below **BRISTOL BATH ROAD SHED – NO. 6957**
Over the ash pits at Bristol Bath Road shed is Collett 'Hall' Class no. 6957 *Norcliffe Hall*. The engine has Bristol St Philip's Marsh '82B' shed code here which was present from October 1959 to February 1960. Photograph by Bill Reed.

Above **BRISTOL BARROW ROAD SHED – NO. 44269 AND NO. 62057**
A particularly unusual pairing has been caught at Bristol Barrow Road shed. Fowler 4F Class 0-6-0 no. 44269 (a resident) is next to Peppercorn K1 2-6-0 no. 62057 from York. In the late 1950s, the locomotive was recorded making the uncommon journey on a goods duty. Photograph courtesy Colour-Rail.

Below **BRISTOL ST PHILIP'S STATION**
In 1870, the Midland Railway opened the small St Philip's station to relieve pressure on Temple Meads. Mainly seeing local passenger trains, these services were withdrawn in 1953 and goods were accepted to 1967 when closed. The building is seen on 16th June 1962. Photograph by B.W.L. Brooksbank.

Above **BRISTOL BATH ROAD SHED – NO. 7019**
At the south end of Bristol Temple Meads station, the Bristol & Exeter Railway established locomotive servicing facilities in 1850. The building possessed six lines and was later extended to increase capacity. Following the takeover by the GWR in 1876, the company converted a second building for stabling and this was a double roundhouse. The latter was modernised by the GWR in the mid-1930s and became a straight shed with ten tracks. This was in use to 1960 when the site was repurposed for diesel locomotives. Collett 'Castle' Class 4-6-0 no. 7019 *Fowey Castle* stands in the yard in 1959/early 1960. In 1958, the locomotive saw the addition of a double chimney. Photograph by Bill Reed.

Opposite above **BURNHAM-ON-SEA – NO. 41202**
On 28th April 1957, the Railway Correspondence & Travel Society's 'North Somerset' railtour has reached Burnham-on-Sea on the short branch from Highbridge. The station was opened by the Somerset Central Railway in 1858 as Burnham, with the change occurring in 1920 under the S&DJR. Thirty-one years later, the station closed for regular trains, but still accepted excursions and freight to 1963. The railtour started at Waterloo with 'King Arthur' Class no. 30453 *King Arthur* and travelled to Reading where recently restored Churchward 3700 Class no. 3440 *City of Truro* hauled the party to Bristol via Savernake and Bath. A pair of Ivatt 2MTs was then engaged to take the train around North Somerset, though only no. 41202 was used to Burnham-on-Sea and no. 41203 remained at Highbridge. From Bristol, Collett 4575 no. 5528 assisted no. 3440 with the railtour which terminated at Paddington. Photograph courtesy Rail-Online.

Opposite below **BRISTOL TEMPLE MEADS STATION – NO. 5061**
In the 1890s, the GWR improved the Paddington to Penzance service with a fast express which was unofficially titled the 'Cornishman'. This name was lost in the early 20th century as the service evolved but in 1952 the 'Cornishman' was revived for a Wolverhampton-Penzance train. The train ran from both ends, leaving the north at 09.15 and the south at 10.30. A stop at Bristol occurred at 12.28 and 15.56 respectively and lasted several minutes. Collett 'Castle' Class no. 5061 *Earl of Birkenhead* has paused at Bristol Temple Meads station on 29th August 1960. During the decade the service was moved on to the ex-Midland line and extended to Derby, Sheffield and Bradford. Photograph by Bill Reed.

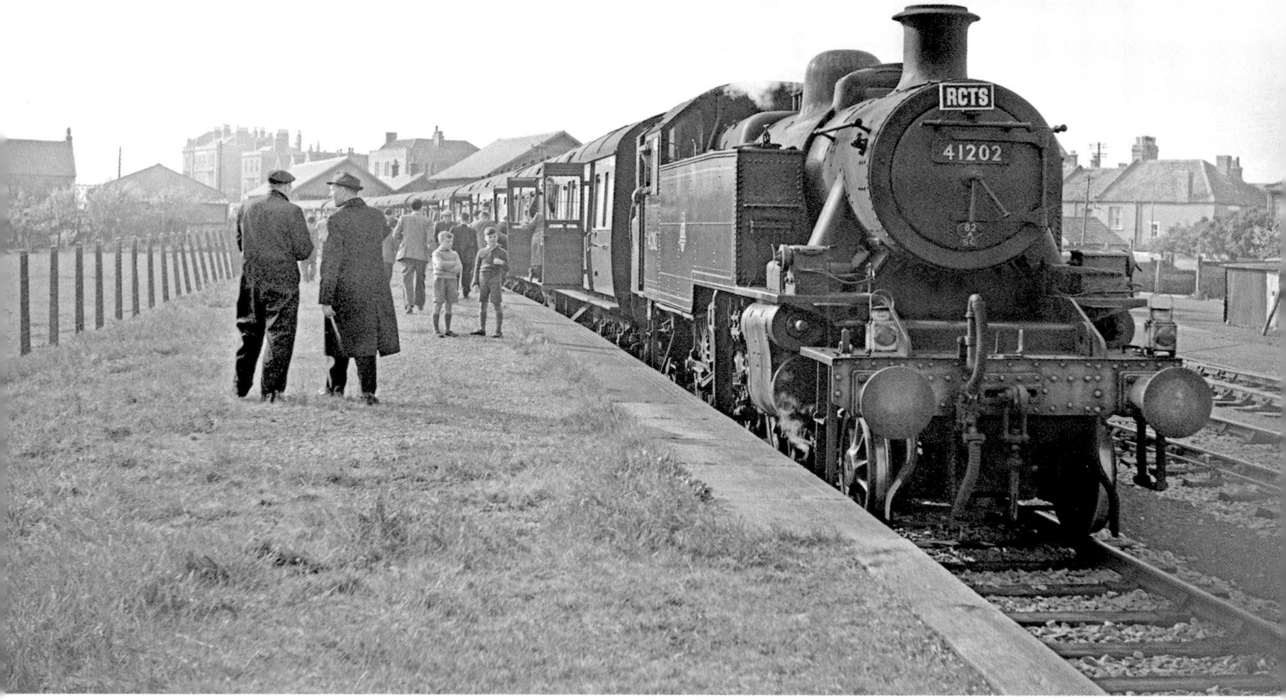
RCTS
41202

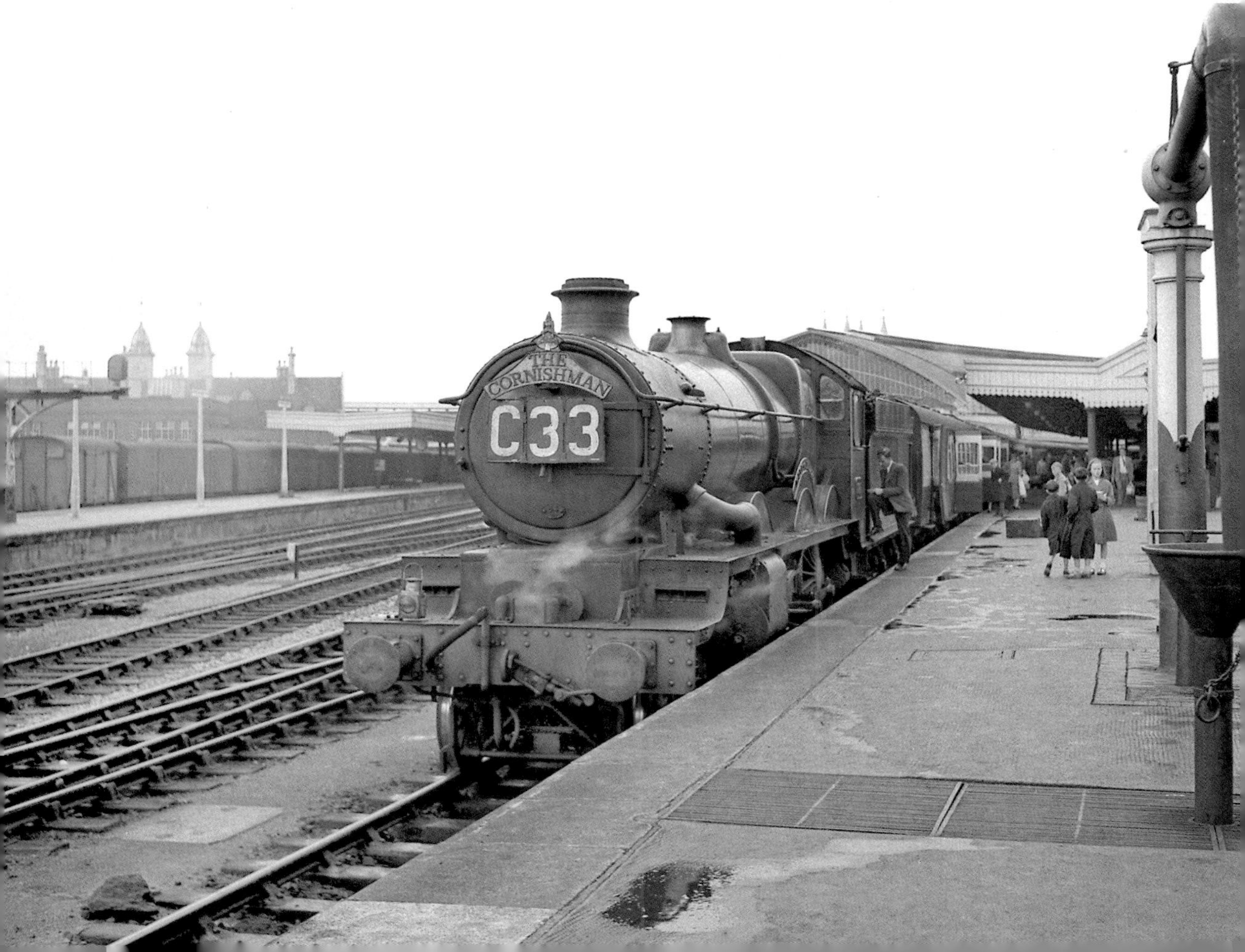
THE CORNISHMAN
C33

Above **BUDE STATION – NO. 31859**
The Devon & Cornwall Railway between Plymouth and Exeter provided the springboard for a branch to Holsworthy in 1879. The ten miles westward to Bude were travelled by coach at this time as the L&SWR declined providing a rail extension. Before the turn of the century this had been laid owing to Bude's development as a resort and this was the main traffic to closure in 1966. Freight had ceased to be handled at the station two years earlier. Two months before this, on 7th July 1964, Maunsell N Class 2-6-0 no. 31859 has the 13.15 goods service from Bude to Exmouth Junction. Photograph by Dave Cobbe courtesy Rail Photoprints.

Opposite above **CHALFORD STATION – NO. 9453**
The Cheltenham & Great Western Union Railway's line existed from the mid-1840s, running between Swindon, Kemble, Stonehouse, Gloucester and Cheltenham. The route skirted the southern edge of the village of Chalford, yet a station was not built for residents until 1897. Hawksworth 9400 Class 0-6-0PT no. 9453 arrives at Chalford with a local service from Gloucester on 26th September 1964. Just over a month remained for the station as closure occurred on 2nd November. The locomotive was also withdrawn just a few days later. Photograph by David Christie.

Opposite below **CHALFORD STATION – NO. 6863**
From Brimscombe, the line rose for five miles at gradients including 1 in 60, 1 in 70, 1 in 90 and 1 in 100. Chalford station was situated around two miles into the climb which was on the way to Swindon. In the opposite direction, the summit was reached after just two or three miles at 1 in 90, 1 in 94 and 1 in 100. Collett 'Grange' Class 4-6-0 no. 6863 *Dolhywel Grange* has a train of mineral hopper wagons at Chalford station on 26th September 1964 and is being banked at the rear by Collett 6100 Class 2-6-2T no. 6106. Photograph by David Christie.

Above **CHALFORD STATION – NO. 1455**
Collett 1400 Class 0-4-2T no. 1455 has made a stop at Chalford station on 23rd July 1963. The engine had recently joined the ranks at Gloucester following a move from Banbury. Photograph by Geoff Warnes.

Below **CHARD JUNCTION – NO. 30841**
An Exeter Central to Yeovil Town local train is at Chard Junction with Urie S15 Class 4-6-0 no. 30841. The locomotive was built in July 1936 and had a service life that lasted to January 1964. Photograph by Bill Reed.

Above **CHARD CENTRAL STATION – NO. 9670**
The passenger service at Chard Central was withdrawn in 1962. Collett 8750 Class no. 9670 has a local train to Chard Junction, c. 1960. Photograph by Bill Reed.

Below **CHARD JUNCTION STATION – NO. 9663**
The Exeter-Yeovil route passed to the south of Chard in 1860. A station – Chard Road – was provided for passengers to continue the three-mile journey. Then, a short branch was built in 1863 and used up to 1966. Collett 8750 Class no. 9663 is ready to make the journey, c. 1960. Photograph by Bill Reed.

Above **CHELTENHAM SPA MALVERN ROAD STATION – NO. 2241**
Cheltenham's first station was built for traffic on the Cheltenham & Great Western Union Railway and in use from 1847. In the early 1890s, a new station was built on a slightly different site. This was partially problematic being a terminus as through traffic had to reverse to continue journeys. To ease this obstacle, a new station was constructed to the south west – Cheltenham Spa Malvern Road – and opened in 1908. A lively floral arrangement contrasts with a grime-covered locomotive working a stopping freight on 22nd June 1963. The offending engine is Collett 2251 Class 0-6-0 no. 2241. Employed at Hereford, the locomotive was there in February 1964 when condemned. Malvern Road station survived a further two years and the site is now covered by housing. Photograph by Geoff Warnes.

Opposite above **CHIPPENHAM STATION – NO. 1462**
Swindon's 1400 Class 0-4-2T no. 1462 has a short train at Chippenham station during April 1958. In the background is the Westinghouse Brake & Signal Company building. This was built in the 1890s and the company had a presence in the town for a number of years. The name has been lost subsequently following a takeover by Siemens. No. 1462 departed Swindon for Exeter in March 1959 and left traffic there in September 1962. Photograph by Bill Reed.

Opposite below **CHIPPENHAM STATION – NO. 5044**
Collett's 'Castle' Class 4-6-0 design was built from just after Grouping to 1950 when 171 examples were in traffic. Gaps of several years existed between batches and no. 5044 was the second engine built following a pause of three years from 1933 to 1936. Those built as part of this order were initially named after castles, though in 1937 they received the titles of Earls. At first, these had been allocated to 3200 Class 4-4-0s in an apparent joke on peers requesting engines to be named after them but this was soon reversed. No. 5044 was in traffic from March 1936 to September 1937 as *Beverston Castle* but was then known as *Earl of Dunraven* which was in the Irish Peerage, though the family had connections to South Wales. No. 5044 is at Chippenham Station during April 1958. Photograph by Bill Reed.

Above **CHURSTON STATION – NO. 1466**
The railway expanded slowly southward from the main line at Newton Abbot. The South Devon Railway opened a branch to Torquay in 1864. Originally, this route was to extend to Paignton, Kingswear and Brixham but was dropped for several reasons. The Dartmouth & Torbay Railway took the project and reached Paignton in 1859, then to Brixham Road (renamed Churston) in 1861 and Kingswear in 1864. Residents and businesses of Brixham were unhappy that the rail connection was two miles away resulting in the formation of the Torbay & Brixham Railway in 1864, with opening occurring during 1868. The whole route was operated by the SDR until the GWR took over in 1876. The Brixham branch closed in 1963, whilst Churston survived to 1967. The line from Paignton to Kingswear was later preserved as the Dart Valley Railway. A Brixham branch train is in the bay at Churston, c. 1960, with 1400 Class no. 1466. Photograph by Bill Reed.

Opposite above **CHRISTOW STATION – NO. 5536**
In the early 1860s, two lines were authorised to branch from the South Devon Railway's line between Exeter and Plymouth. One left the route north of Newton Abbot and reached Moretonhampstead in 1866. The other was not ready until 1882 and made a junction at Heathfield on the Moretonhampstead branch. The route then followed the Teign Valley to rejoin the main line near Exeter. Christow station only opened in 1903 and provided the solitary passing loop on the Teign Valley Railway. With declining passenger numbers, the line was marked for closure in June 1958 and this image is from the last day, featuring a number of people marking the demise. Collett 4575 Class no. 5536 has one of the final trains at Christow station. Freight lingered for a time before flooding damaged the trackbed in the early 1960s. Photograph by L.F. Folkard courtesy Colour-Rail.

Opposite below **CHURSTON STATION – NO. 1452**
George Armstrong's 517 Class 0-4-2T locomotives were built for the GWR's branch-line traffic from 1868 to 1885. Numbering 156, these needed replacement by the 1930s and Collett produced the 4800 Class 0-4-2T, later reclassified 1400. Appearing in 1932, 75 engines were in service by 1936. No. 1452 was built at Swindon in July 1935 and had a service life of nearly 25 years. The locomotive is in the bay platform at Churston station on 4th August 1958 with a Brixham branch service. Newton Abbot-allocated at the time, withdrawal from Exeter occurred in June 1960. Photograph by Bill Reed.

Above **CROWCOMBE STATION – NO. 2240**
Collett's mixed traffic 0-6-0 2251 Class appeared steadily from 1930 to Nationalisation when 120 engines had been erected. No. 2240 was one of the late examples and ready for service in December 1944. When pictured on 12th August 1961, no. 2240 was 16 years old and employed at Taunton. One of the depot's duties has taken the engine to the north west on the Minehead branch and the train has reached Crowcombe station. The West Somerset Railway established the facilities in 1862 as Crowcombe Heathfield, though by 1890 just Crowcombe was used. The station survived past the Beeching closures to 1971 and was later resurrected as part of the West Somerset Railway which at present has 20 miles of line between Minehead and Bishop's Lydeard. Photograph by Bill Reed.

Opposite above **CORFE CASTLE STATION – NO. 41303**
William the Conqueror built over 30 castles following his invasion of England. Corfe Castle was one of these, being located in a Royal Forest that was favoured by King William for hunting. Sitting between two hills, the castle was built from stone which was a distinguishing feature for the time. A number of following monarchs made additions and improvements to the fortress, though by the reign of Elizabeth I had lost some importance and she sold the castle to Lord Chancellor Sir Christopher Hatton. During the English Civil War, Corfe Castle was held by Sir John Bankes, Chief Justice, and his wife defended against Parliamentarian forces for several weeks. Ultimately infiltrated, Corfe Castle was later partially demolished. The remains provide a dramatic backdrop here on 5th June 1964 as Ivatt 2MT no. 41303 approaches Corfe Castle station with a train for Swanage. The fireman has the token ready for exchange. Photograph by David Christie.

Opposite below **CORFE CASTLE STATION – NO. 30057**
As the line to Dorchester skirted the top of the Isle of Purbeck peninsula, thoughts soon turned to providing a line to Swanage for improved movement of products derived from the area. A number of plans were formulated but only solidified in the early 1880s. The Swanage Railway ran from Wareham to Corfe Castle and Swanage, opening in 1885; the L&SWR took over in the following year. Before closure in the 1970s, traffic had transitioned to consist of tourists. Like the West Somerset Railway, the Swanage Railway was formed by the end of the decade to operate as a heritage line. Drummond M7 Class no. 30057 has a passenger service at Corfe Castle station, c. 1960. Photograph by Bill Reed.

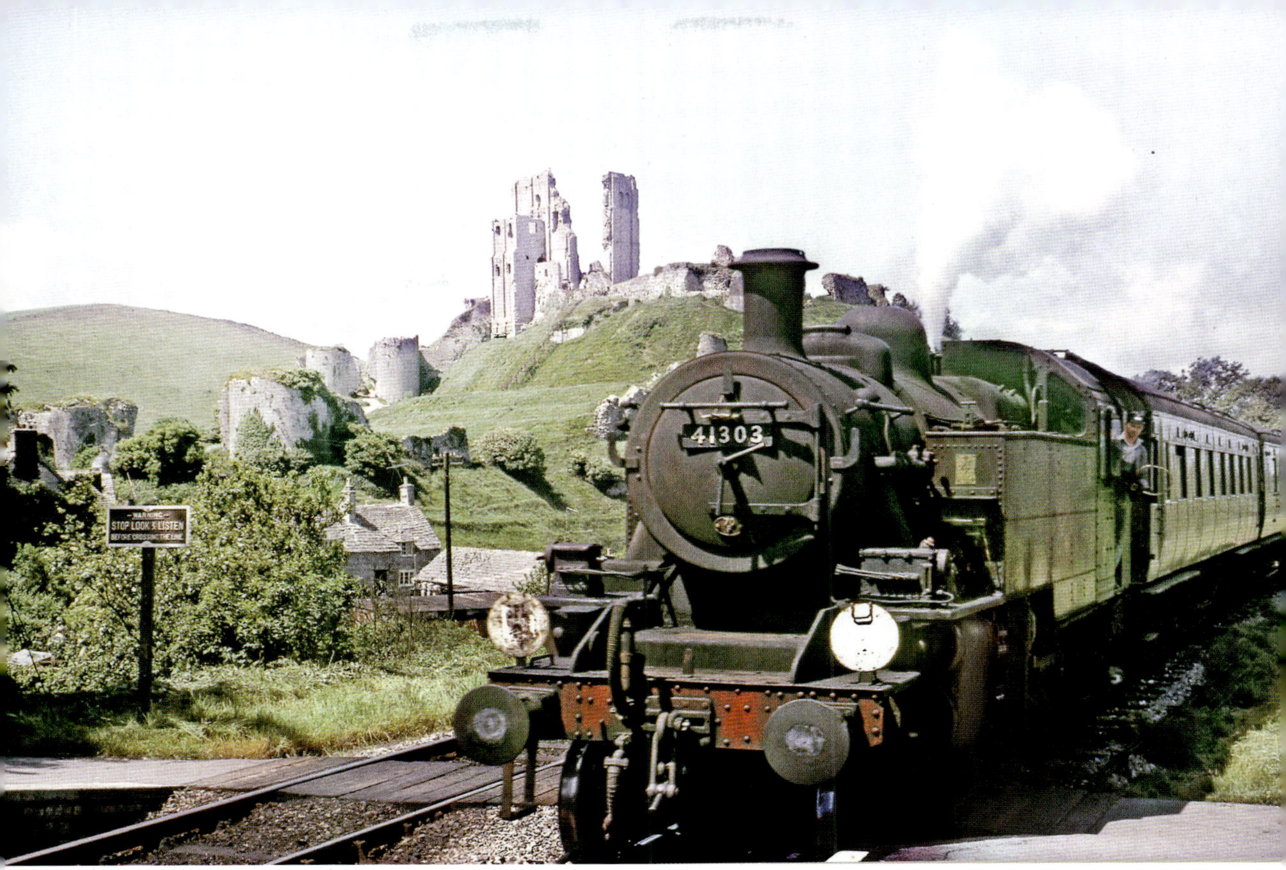
41303

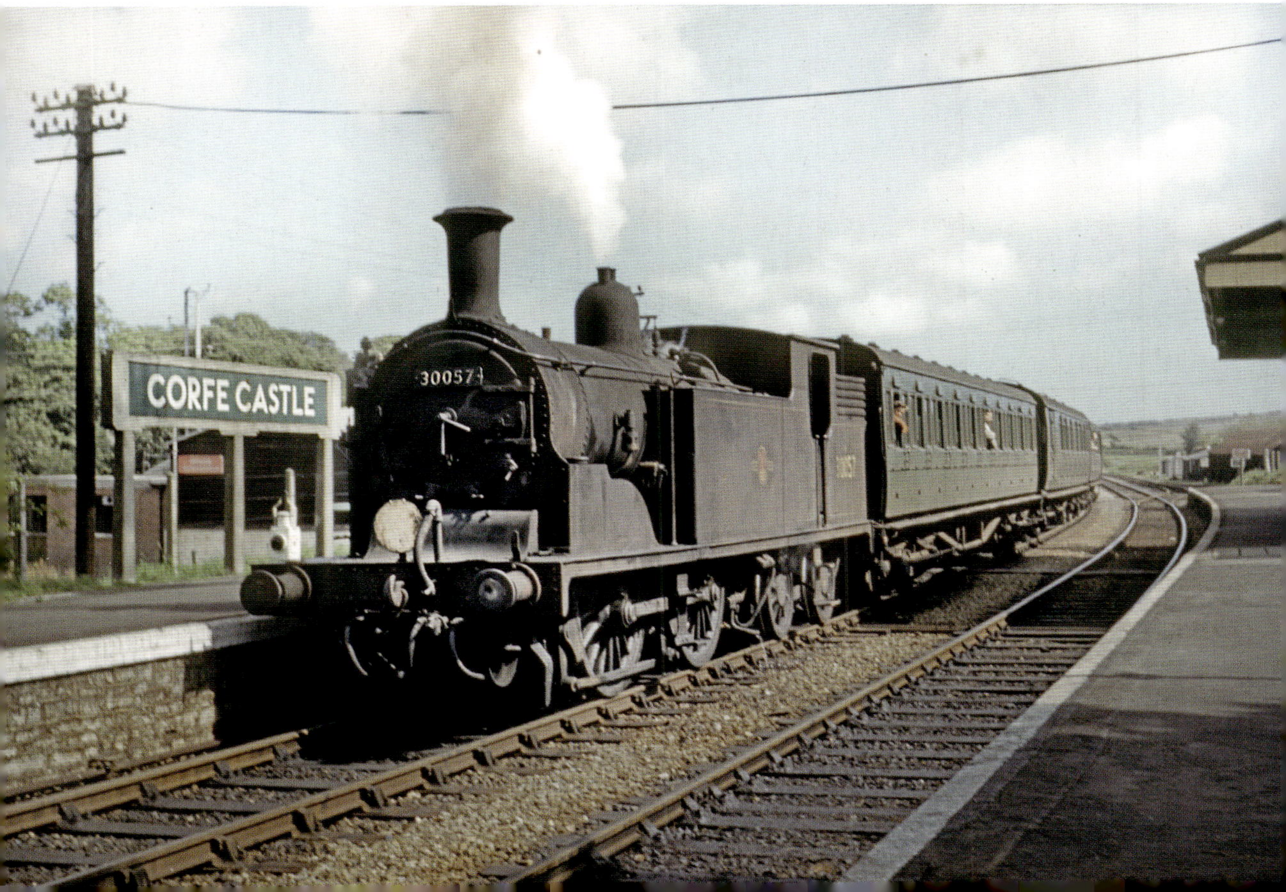
CORFE CASTLE
30057

Above DAINTON BANK – NO. 70024 AND 6018
To the south west of Newton Abbot, Dainton bank provided a particular challenge for trains as around three miles of gradients near 1 in 60 existed. BR 'Britannia' Class Pacific no. 70024 *Vulcan* is piloting Collett 'King' Class no. 6018 *King Henry VI* on 27th July 1957. Photograph by B.W.L. Brooksbank.

Below DAWLISH – NO. 6385
An Exeter to Kingswear local train passes promenaders on the sea front at Dawlish on 26th July 1958. The locomotive is Churchward 4300 Class no. 6385. Photograph by B.W.L. Brooksbank.

Above DAWLISH STATION – NO. 5339
A local train is at Dawlish station on 5th September 1954 with Churchward 4300 Class no. 5339. Prominent on the left is the signal box which was built in 1920 and in use for five decades. The box stood derelict for a number of years until demolished in 2013. Photograph by Geoff Warnes.

Below DEVONPORT KINGS ROAD STATION – NO. 34024
An eastbound train to Exeter is ready to leave Devonport Kings Road station with rebuilt 'West Country' Pacific no. 34024 *Tamar Valley*. Seen on 18th August 1961, the engine had been rebuilt just six months earlier. Photograph by Dave Cobbe courtesy Rail Photoprints.

Opposite above **DULVERTON STATION – NO. 7337**
The Devon & Somerset Railway completed their route of 43 miles between Taunton and Barnstaple in 1873. Dulverton station opened at this time, but was a short distance away from the town. Two platforms were provided, along with goods shed and station master's house. The signal box was originally on the opposite platform, then in 1908 moved to the position here. The branch to Tiverton opened in the mid-1880s and soon extended to Exeter. Trains on this line were accommodated on the south side of the westbound platform, or the left of the camera position here. Collett 5700 Class 2-6-0 no. 7337 has a short freight Taunton-bound, c. 1960. Photograph by Bill Reed.

Opposite below **DUNMERE HALT STATION – NO. 30586**
The 09.35 Wadebridge to Wenford Bridge freight passes Dunmere Halt station with Beattie 0298 Class 2-4-0WT no. 30586. Pictured on 16th May 1962, the locomotive had just six months left in service. Photograph by Dave Cobbe courtesy Rail Photoprints.

Below **DULVERTON STATION – NO. 9765**
A batch of 25 8750 0-6-0PT locomotives was produced at Swindon between September 1935 and July 1936. No. 9765 was amongst this number and was ready for service in late September 1935. Operational for just over 26 years, for the last eight years in traffic, the engine's allocation was to Exeter. No. 9765 has likely travelled via Tiverton to Dulverton with a local train, c. 1960. Photograph by Bill Reed.

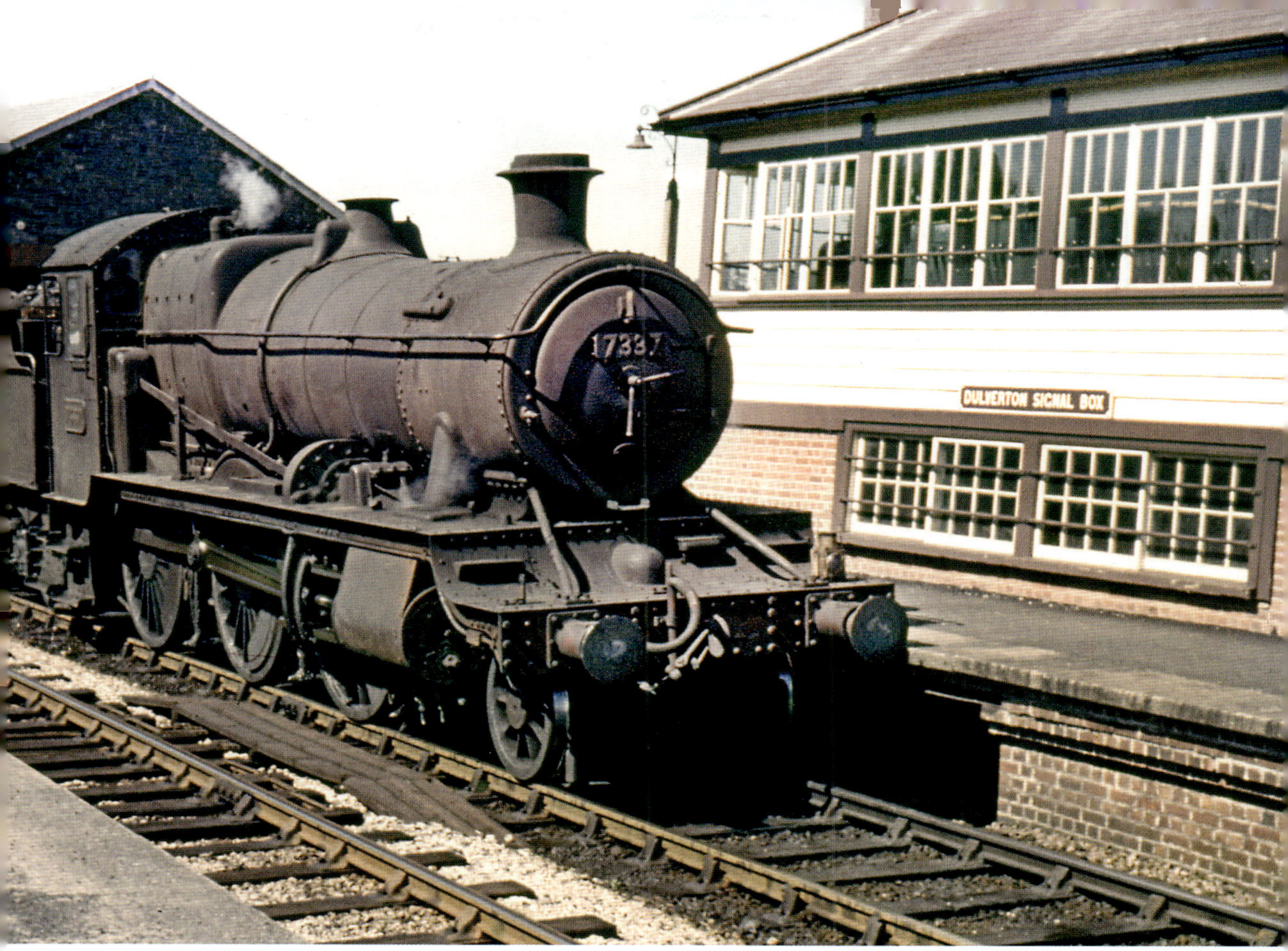
17337
DULVERTON SIGNAL BOX
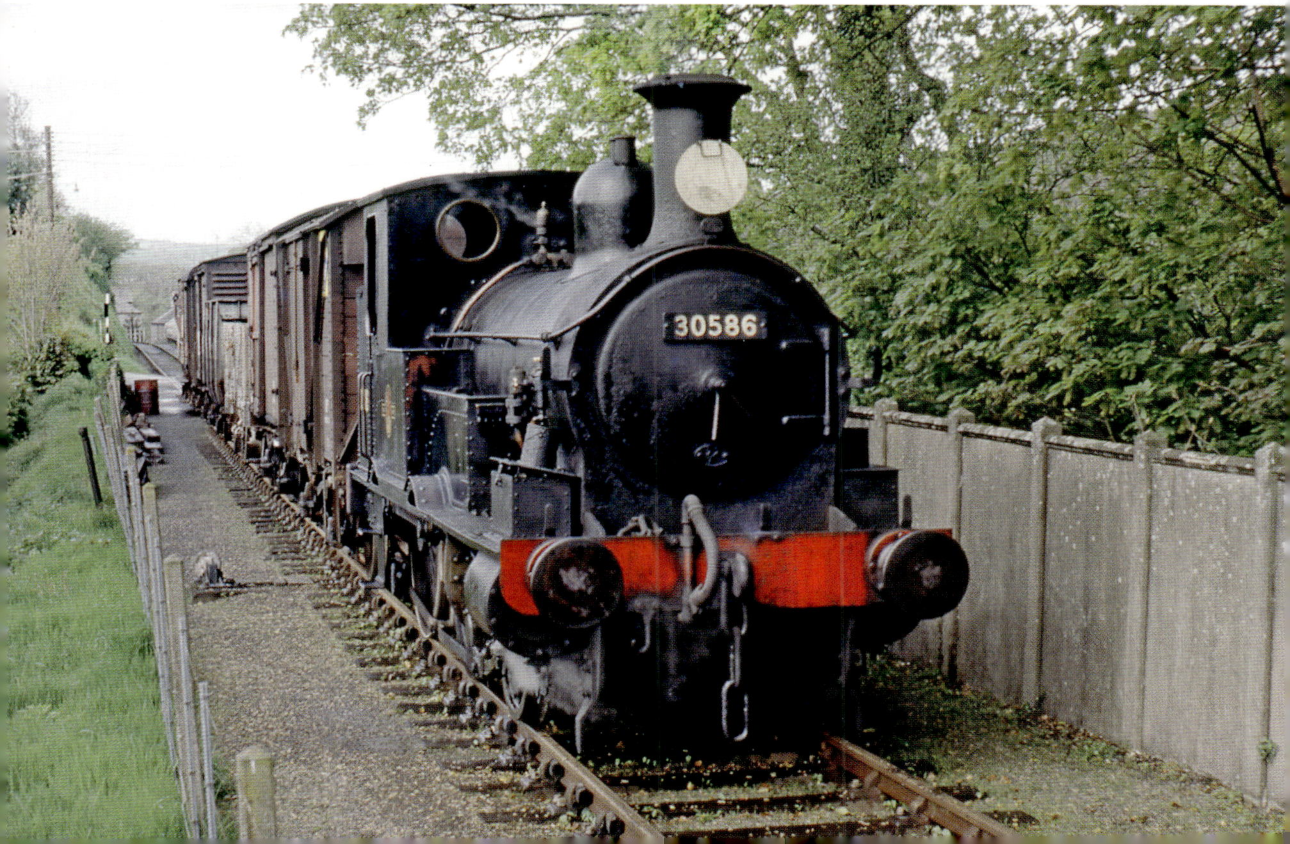
30586

Above **EXETER ST DAVIDS STATION – NO. 34029**
Although the London & South Western Railway opened their own station at Exeter in 1860 (Queen Street, later Central), a short connection was made to St Davids in 1862. This also provided a link to the Exeter & Crediton Railway which was a part of the L&SWR's westward expansion. In the late 1950s, Bulleid 'West Country' Pacific no. 34029 *Lundy* has a train at St Davids station. The engine has an experimental livery where the waistline of the casing and tender were lined and the area below was painted black. In December 1958, no. 34029 was rebuilt and continued working to September 1964. Allocated to Exmouth Junction when pictured, the locomotive transferred to Nine Elms, then Bournemouth in 1959, with the latter move being the final one. Photograph by Bill Reed.

Opposite above **EVERCREECH JUNCTION STATION – NO. 53809 AND NO. 34103**
The Somerset & Dorset Joint Railway had locomotives provided by the Midland Railway and coaching stock from the London & South Western Railway. In the early 20th century, the line called for a powerful engine capable of working the difficult sections within certain permanent way limits. The 7F Class 2-8-0 was the result and a total of 11 appeared in two batches – six in 1914 and five in 1925. Whilst Derby built the first lot, the second was acquired from Robert Stephenson & Co. and no. 53809 was erected there in July. The locomotive is piloting 'West Country' Class Pacific no. 34103 *Calstock* on 1st September 1962. The pair are at Evercreech Junction station with a northbound Saturday train from Bournemouth. Photograph from the Dave Cobbe Collection courtesy Rail Photoprints.

Opposite below **EVERCREECH JUNCTION STATION – NO. 80041**
Evercreech Junction was built as Evercreech in 1862 on the Somerset Central Railway's extension from Glastonbury to Cole. Evercreech was later chosen as the junction for the Bath Extension in 1874. The station was open to 1966. BR Standard Class 4MT 2-6-4T no. 80041 is taking water at Evercreech on 3rd July 1965. Withdrawal occurred in March the following year at the demise of the S&DJR. Photograph by David Christie.

Above **EXETER ST DAVIDS STATION – NO. 5050**
Collett 'Castle' Class no. 5050 *Earl of St Germans* is in the process of taking water at Exeter St Davids station in the early 1960s. The engine had been paired with a Hawksworth tender for the late 1950s, but has returned to the Collett-type here. Photograph by Bill Reed.

Below **EXETER ST DAVIDS STATION – NO. 5049**
No. 5049 *Earl of Plymouth* had a Hawksworth tender like classmate no. 5050 (see above) in the late 1950s and changed for the new decade. The type is still behind the engine at Exeter St Davids station. Photograph by Bill Reed.

Above **EXETER ST DAVIDS STATION – NO. 6938**
Originating with the Bristol & Exeter Railway in 1844, Exeter St Davids station, which was designed by Isambard Kingdom Brunel, welcomed the South Devon Railway as well in 1846. With the arrival of the L&SWR, the station was partially rebuilt and again just before the First World War. Collett 'Hall' Class no. 6938 *Corndean Hall* has a local train c. 1960. Photograph by Bill Reed.

Below **EXETER ST DAVIDS STATION – NO. 7224**
No. 7224 started life as a Collett 5205 Class 2-8-0T in 1926 but around ten years later was converted to a 2-8-2T becoming a 7200 Class member. The locomotive was only briefly at Exeter, covering the period June to December 1961, and mainly resided in South Wales. Photograph by Bill Reed.

Opposite above **EXMOUTH STATION – NO. 30024**
Drummond M7 Class no. 30024 is coupled to an Ivatt Class 2MT 2-6-2T, with the number appearing to be 41306, at Exmouth station. No. 30024 was a long-term servant at Exmouth Junction shed, whilst no. 41306 arrived in June 1955 and remained until condemned at the end of 1963. No. 30024 departed for Bournemouth in late 1962, though was sent for scrap during March 1963. Photograph by Bill Reed.

Opposite below **EXMOUTH JUNCTION SHED – NO. 30689**
On 24th February 1963, Drummond 700 Class 0-6-0 no. 30689 is fitted with a much-needed snow plough at Exmouth Junction shed. The 'Big Freeze' had started in late December and there were a number of periods of drifting snow up to 20 ft deep in places. Photograph by Bill Reed.

Below **EXMOUTH JUNCTION SHED – NO. 30451**
The First World War started as Robert Urie was planning a new express locomotive for the L&SWR. He had recently introduced the H15 Class 4-6-0 for mixed traffic duties and early experience suggested the design was suitable for development to meet the passenger need. Towards the end of the conflict, the first ten N15 Class appeared, followed by another ten before Grouping. Richard Maunsell became Chief Mechanical Engineer of the Southern Railway and decided to build another ten before introducing his own express locomotive. These saw some improvements to the design, including the motion, superheater and draughting. No. 30451 *Sir Lamorak* was one of these engines, being constructed at Eastleigh Works in June 1925. The engine is at Exmouth Junction shed, c. 1960. From 1950 until condemned, no. 30451 was employed at Salisbury depot. Photograph by Bill Reed.

Above **FALMOUTH STATION – NO. 5537**
A local service from Truro has reached Falmouth station in the late 1950s. Collett 4575 Class no. 5537 is at the head of the train and was a long-term servant at Truro. From early 1962, the locomotive had several months at Penzance before condemned during August. Photograph by Bill Reed.

Opposite above **EXMOUTH STATION – NO. 30323**
Though the railway had been at Exeter since the early 1840s, interested parties in a line to Exmouth did not favour a broad-gauge railway. As a result, the promoters waited for the arrival of the L&SWR and standard gauge. This happened towards the end of the 1850s as a line from Yeovil to Exeter was built, in addition to half of the 11-mile branch to Exmouth. The project was fulfilled in 1861. The original station at Exmouth was rebuilt in the 1920s, and again in 1976. In April 1958, Drummond M7 Class no. 30323 takes water at the station. The locomotive possessed a 1,300-gallon tank capacity. Photograph by Bill Reed.

Opposite below **FALMOUTH STATION – NO. 4574**
The Cornwall Railway originally planned to run from Plymouth to Falmouth. Yet, encountering financial problems, their route only reached Truro in 1859. The CR had to enter into an agreement with other local companies to raise capital which allowed the remainder of the line to be laid. The first trains ran in 1863 and terminated at Falmouth which had a three-platform station with goods and engine sheds also provided. Collett 4500 Class no. 4574 is with a train at Falmouth on 30th August 1958. Twelve years later the station closed and was replaced by Falmouth Town, only to reopen in 1975 now known as Falmouth Docks. Photograph by L. Rowe courtesy Colour-Rail.

Above **FOWEY STATION – NO. 1408**
A local train to Lostwithiel is ready to leave Fowey station with Collett 1400 Class no. 1408 during June 1956. A year earlier, the locomotive had moved to Plymouth. Withdrawal took place there in March 1958. Photograph courtesy Rail-Online.

Below **FROME SHED – NO. 4555**
No. 4555 was the first locomotive of a batch of 20 engines built to Churchward's 4500 Class design and these were the last before the introduction of Collett's 4575 Class. The locomotive is at Frome shed during April 1958 and was allocated to the nearby Westbury shed at the time. Photograph by Bill Reed.

Above **FROME – NO. 3735**
A mineral train is running on the line from Radstock to Frome with 8750 Class no. 3735. Pictured on 3rd April 1965, the locomotive was withdrawn in September. Photograph by Hugh Ballantyne courtesy Rail Photoprints.

Below **FROME SHED – NO. 9615**
Servicing facilities opened just to the south of Frome station, on the west side, in the mid-1840s. This was rebuilt at the end of the century to the one-track timber building which has 8750 Class no. 9615 in front, c. 1960. Westbury-allocated, the engine moved on to Bristol at the end of 1961. Photograph by Bill Reed.

Above **GARA BRIDGE STATION – NO. 5558**
The 11.15 Kingsbridge to Brent local train accelerates away from Gara Bridge station on 6th August 1960. The engine is 4575 Class no. 5558 which was built in November 1928 and operational until October 1960. Photograph by Hugh Ballantyne courtesy Rail Photoprints.

Below **GLOUCESTER CENTRAL STATION – NO. 4135**
Collett 5101 Class 2-6-2T no. 4135 eagerly awaits departure from Gloucester Central station on 5th September 1962. Photograph by Bill Reed.

Above **GLASTONBURY & STREET STATION – NO. 41296**
The 13.15 from Evercreech Junction to Highbridge has paused for passengers at Glastonbury & Street station. Ivatt Class 2MT no. 41296 leads the train on 11th December 1965. Photograph by Hugh Ballantyne courtesy Rail Photoprints.

Below **GLOUCESTER EASTGATE STATION – NO. 75022**
The ex-Midland Railway station at Gloucester became Gloucester Eastgate in 1951 and used this title to closure in 1975. A local passenger train from Bristol has made a stop on 13th April 1959, with BR Standard Class 4 no. 75022 in charge. Photograph by B.W.L. Brooksbank.

Above GLOUCESTER EASTGATE STATION – NO. 7029

The Stephenson Locomotive Society organised a special featuring Bulleid's Pacifics on 23rd May 1965. No. 34051 *Winston Churchill* took the train from Birmingham to Salisbury and no. 35017 *Belgian Marine* was at the head from there to Westbury. At the latter 'Castle' Class no. 7029 *Clun Castle* was used to take the party back to Birmingham. A stop is taking place at Gloucester Eastgate for the crew to change and replenish the water supply. Photograph courtesy Rail-Online.

Above **GLOUCESTER SHED – NO. 5068**
Collett 'Castle' Class no. 5068 *Beverston Castle* is at Gloucester shed for servicing. Fitted with a double chimney, the image dates post-March 1961. The engine was condemned in September 1962. Photograph by Bill Reed.

Below **GLOUCESTER CENTRAL STATION – NO. 5026**
Under preparation to continue with a goods train, 'Castle' no. 5026 *Criccieth Castle* has stopped at Gloucester Central station on 26th September 1964. Photograph by David Christie.

GLOUCESTER SHED – NO. 4929

Particularly well presented, 'Hall' Class no. 4929 *Goytrey Hall* is in the yard at Gloucester. The engine was a long-term resident. Photograph by Bill Reed.

Above **GLOUCESTER SHED – NO. 5955**
On 11th September 1952, 'Hall' Class no. 5955 *Garth Hall* has worked to Gloucester from Landore, South Wales, and is serviced before returning. Photograph by B.W.L. Brooksbank.

Below **GLOUCESTER CENTRAL STATION – NO. 1630**
Hawksworth 1600 Class 0-6-0PT no. 1630 has a local train at Gloucester Central station. The engine was a late addition to stock in January 1951 and had a service life that lasted to June 1964. Photograph by Bill Reed.

GWINEAR ROAD STATION – NO. 6911
'Hall' Class no. 6911 *Holker Hall* speeds through Gwinear Road station with an express during July 1956. Photograph by Bill Reed.

Above **GWINEAR ROAD STATION – NO. 4548**
Gwinear Road station was the point of departure for passengers to Helston, an important town in the area. This was connected to the main line between 1887 and 1964. A branch train is with Churchward 4500 Class no. 4548 during July 1956. Photograph by Bill Reed.

Below **GWINEAR ROAD STATION – NO. 7916**
An eastbound express approaches the platform at Gwinear Road station on 22nd June 1956. The engine is Newton Abbot's Hawksworth 'Modified Hall' Class no. 7916 *Mobberley Hall*. Gwinear Road closed with the branch in 1964. Photograph by G.H. Hunt courtesy Colour-Rail.

Above **HALWILL STATION – NO. 30715**
An eastbound passenger train, with goods vans at the rear, has paused at Halwill station. Drummond T9 Class 4-4-0 no. 30715 is leading. Under BR, the engine was based at Exmouth Junction shed. Photograph by Bill Reed.

Below **HALWILL STATION – NO. 41295**
A number of sidings existed at Halwill for the transfer of freight between four routes. A goods train is seen in the distance while another stands off to the left. A short local passenger train is on the right with Ivatt 2MT no. 41295 of Barnstaple. Photograph by Bill Reed.

Above **HALWILL STATION – NO. 31853**
Opened in 1879, Halwill station was initially on the L&SWR line to Holsworthy, ultimately extended to Bude. Then, a branch southward reached the North Cornwall Railway at Launceston, while a route northward to Torrington, Barnstaple and Ilfracombe came into use. Maunsell N Class no. 31853 is at Halwill station, c. 1960. With '72A' on the smokebox door, the engine had worked from Exmouth Junction. Photograph by Bill Reed.

Below **HALWILL STATION – NO. 82019**
Another short passenger train approaches Halwill station, though in this instance is BR Standard Class 3 2-6-2T no. 82019. The engine went on to be one of the last class members in service. Photograph by Bill Reed.

Below **HENSTRIDGE STATION – NO. 75071**
The last batch of 15 BR Standard Class 4 4-6-0s were all allocated for duties on the Southern Region. As a result, these engines were paired with a BR1B tender which had a higher water capacity than others used with the Class 4s. This was due to the lack of sufficient water troughs in the area. No. 75071 was new from Swindon Works in October 1955. The locomotive transferred to Bath Green Park in June 1956 and remained on the ex-S&DJR to 1964, with the last two years of this spent at Templecombe. In 1958 the northern half of the route transferred to the Western Region. Several Class 4s were used on the line for local trains and this signalled the demise of 2P Class 4-4-0s. No. 75071 has a local train south of Templecombe at Henstridge station, c. 1960. Photograph courtesy Rail-Online.

Above HELSTON STATION – NO. 4505

The early addition of a railway to Halston and the surrounding area would have been beneficial for locals and businesses. Yet, the difficult terrain discouraged many proposals and not until 1880 did an Act finally receive approval. By this time industrial concerns had been reduced but agricultural traffic was prominent and thankful for the eventual opening in 1887. Helston was the terminus of the line from Gwinear Road and also the most southerly in England. Though the route ended at Helston there were plans to extend further south, but these never came to fruition. Along with the station, a goods shed, locomotive servicing facilities and carriage shed were built. Churchward 4500 Class no. 4505 is reversing back to connect with the 17.20 to Gwinear Road on 24th July 1957. The engine was only in service for another three months before condemned. The Helston branch closed in 1964 and lay dormant to the early 2000s when plans for a heritage route were formulated. At present around a mile of line has been resurrected with a selection of motive power and rolling stock in use. Parts of Helston station survive in private ownership. Photograph by P.W. Gray from Rail Archive Stephenson courtesy Rail-Online.

Above **HIGHBRIDGE STATION – NO. 6968**
Collett's 4900 'Hall' Class 4-6-0 was built in numbers from 1924 to 1943. Used for mixed traffic duties, 253 entered service to a generally similar specification. When Frederick Hawksworth became Chief Mechanical Engineer in the early 1940s, he was eager to improve construction practices and features. Welding was used, particularly for tenders and his had flat sides, whilst he changed superheater practice by using more elements. Frame construction was adjusted and cylinders were cast separately from the smokebox saddles. These alterations were embodied first in the 6959 'Modified Hall' 4-6-0. The first was in service during 1944 and 71 appeared to 1950. No. 6968 *Woodcock Hall* was part of the first batch of 12, being erected at Swindon in September 1944. The locomotive has a local train at Highbridge station on 29th August 1960. Earlier in the year, no. 6968 transferred to Westbury and was there nearly two years before moving on to Fishguard. In mid-1963, the engine returned to Westbury and was withdrawn there two months later. Photograph by Bill Reed.

Below HIGHBRIDGE STATION – NO. 3215

Highbridge station opened with the first section of the Bristol & Exeter Railway in 1841. Thirteen years later, the Somerset Central Railway's line connected at Highbridge and ran to Glastonbury. From then on, Highbridge catered for passengers on both lines with two distinct sections. An extension of the SCR ran westward to Burnham-on-Sea from 1858. Collett 2251 Class 0-6-0 is light engine at Highbridge station on 29th August 1960. The locomotive had recently joined the S&DJR at Templecombe and worked there until condemned for scrap in early 1963. Despite the closure of the S&DJR, Highbridge station continues to offer access to the Bristol-Exeter line but has been renamed Highbridge & Burnham station. Photograph by Bill Reed.

Above HIGHBRIDGE STATION – NO. 2244
A parcels train approaches Highbridge station with 2251 Class no. 2244 on 30th August 1960. Photograph by Bill Reed.

Below HONITON STATION – NO. 34057
To the west of Axminster, 'Battle of Britain' Pacific no. 34057 *Biggin Hill* is at Honiton station on 22nd May 1964. Photograph by K.C.H. Fairey courtesy Colour-Rail.

Above **HONITON – NO. 30327**
Honiton bank covered six miles westbound and four miles eastbound on the way to Exeter. The gradients on both sides were mainly below 1 in 100. Drummond 700 Class 0-6-0 is with a breakdown crane in the area during September 1958. Photograph by P.J. Hughes courtesy Colour-Rail.

Below **HAM MILL HALT – NO. 1458**
No. 1458 has a Gloucester to Chalford train near Ham Mill Halt on 26th September 1964. Photograph by David Christie.

Below **KEMBLE STATION – NO. 7010**
The Cheltenham & Great Western Union Railway opened from Swindon to Kemble and Cirencester in 1841. An exchange station was built at Kemble in 1845 and this was only made public from 1st May 1882. Shortly after this date, the GWR planned a branch to Tetbury and this was brought into use by the end of the decade. A westbound train is with resplendent 'Castle' Class no. 7010 *Avondale Castle*, which is perhaps on a running-in turn from Swindon Works; chalked on the cab side is 'arch in'. The engine was in traffic to 1964, at which time both the Cirencester and Tetbury branches were closed. Kemble remains open and has been listed, along with the water tank standing in the background. Photograph by Bill Reed.

Above ILMINSTER STATION – NO. 8783
While the people of Chard were eager for the L&SWR to connect with the company's nearby line, a second project for the Bristol & Exeter Railway to branch southward from near Taunton via Ilminster also found favour. A station at the latter opened on 11th September 1866 with a single platform and second line to allow freight a passing point. Ilminster station was one of many across the country to close during the coal crisis of early 1951, but was reopened by 7th May. The station was closed permanently in 1962. No. 8783 has a train at Ilminster on the last day of the branch. The locomotive was amongst the first batch of 8750 Class 0-6-0PTs which numbered 49 engines. Photograph courtesy Rail-Online.

Another view of 'Castle' no. 7010 *Avondale Castle* at Kemble station on 5th September 1962. Photograph by Bill Reed.

Above **KEMBLE STATION – NO. 5951**
Collett 'Hall' Class no. 5951 *Clyffe Hall* has an express at Kemble station on 5th September 1962. The engine was a long-term employee of Gloucester Horton Road shed. Photograph by Bill Reed.

Below **KEMBLE STATION – NO. 8433**
W.G. Bagnall built Hawksworth 9400 Class 0-6-0PT no. 8433 in March 1953. New to Old Oak Common, the engine transferred to Swindon in 1955. Still employed there when pictured on 5th September 1962 at Kemble station, in 1963 no. 8433 moved back to the original depot. Photograph by Bill Reed.

Above KINGSBRIDGE STATION – NO. 5558

When the South Devon Railway was completed, excitement grew in Kingsbridge for a connection. A local scheme was promoted but failed to generate sufficient funding. In the mid-1860s, a second attempt was made and this led to an authorising Act with capital of £130,000. The project then stalled and could not be restarted until the 1880s when the GWR became involved and agreed to provide most of the financing. The 12-mile route was finally ready at the end of 1893, leaving the main line at Brent and going southward to Avonwick, Gara Bridge and Loddiswell. The terminus was Kingsbridge station which served as the departure point for passengers to the seaside resort at Salcombe. This was the main traffic and ultimately led to the line's closure in September 1963. Collett 4575 Class no. 5558 is light engine at Kingsbridge on 24th July 1958. Photograph by B.W.L. Brooksbank.

Below **KILMERSDON COLLIERY**
The Somerset coalfield covered around 240 sq. miles in the north of the county. Exploitation of outcrops had been taking place from Roman times, but deep mining began during the Industrial Revolution. Kilmersdon colliery was established in 1875 and connected with the Bristol-Frome line which opened two years earlier. The colliery delivered wagons to sidings via a gravity incline. Both steam and rope haulage were used. Peckett 0-4-0ST (works no. 1788) is pictured in January 1969. The locomotive was delivered new to Kilmersdon in 1929 and became one of the last steam engines in the coalfield. When the colliery closed in 1973, the locomotive was preserved with the West Somerset Railway. Photograph by John Chalcraft courtesy Rail Photoprints.

Above LISKEARD STATION – NO. 1007
On 4th May 1859, Liskeard station began serving passengers on the Cornwall Railway line from Plymouth. The facility continues to do so, as well as travellers on the Looe branch which was reprieved from closure in the 1960s. Hawksworth 'County' Class 4-6-0 no. 1007 *County of Brecknock* has a main line express at Liskeard in the mid-1950s. Photograph by Bill Reed.

Opposite LISKEARD STATION – NO. 5539
Mineral traffic was the main commodity transported by the Liskeard & Looe Railway and Liskeard & Caradon Railway from the 1860s. Yet, by the end of the 1870s this was in decline and thoughts turned to passenger traffic. Authorised in September 1879, these trains ran independently of the main line. A connection was finally made with the Cornwall Railway – then GWR – at Liskeard in 1901. A Looe branch train is at Liskeard, c. 1960, with enginemen apparently taking a break on the bench to the right. Photograph by Bill Reed.

Above **LISKEARD STATION – NO. 4569**
Built at Swindon Works in October 1924, Churchward 4500 Class no. 4569 was in traffic for nearly 40 years, being condemned two months away from this milestone. The engine was at Swindon in the last month, but since Nationalisation at least, no. 4569 had spells at several West Country locations. Seen at Liskeard station on 1st September 1960, the locomotive held a St Blazey allocation. Photograph by Bill Reed.

Below **LOSTWITHIEL STATION – NO. 5972**
On 29th July 1958, Collett 'Hall' no. 5972 *Olton Hall* has the 'Cornishman' express eastbound through Lostwithiel station. Photograph by P.W. Gray from Rail Archive Stephenson courtesy Rail-Online.

Above LOOE STATION – NO. 5502

Driver C. Marshall and his fireman pose with Collett 4575 Class no. 5502 at Looe station. The locomotive was erected at Swindon Works in May 1927 and was an early withdrawal in July 1958. The process began for the 4575 Class in 1956 and was completed in 1964. Despite this relatively early demise, 11 class members have survived, many thanks to Woodham Bros Scrapyard. No. 5502 was a long-term resident at St Blazey depot. Photograph by Bill Reed.

LOSTWITHIEL STATION – NO. 1419

A Fowey branch train stands in the bay platform at Lostwithiel station with 1400 Class no. 1419. Photograph by Bill Reed.

Above **LOSTWITHIEL STATION – NO. 8733**
In the 1860s, a line to Fowey from Lostwithiel was promoted mainly for the benefits to china clay traffic. A second competing line was built by the Cornwall Minerals Railway and this ultimately survived and purchased by the GWR. The latter refurbished the line for reopening in 1895 and passenger trains were offered for the first time. These lasted to 1965 but the clay traffic persists today. No. 8733 has this type of train at Lostwithiel. Photograph by Bill Reed.

Below **LYME REGIS STATION – NO. 30584**
A branch train from Axminster is at Lyme Regis station, c. 1960. Adams 415 Class 4-4-2T no. 30584 is the locomotive, which was in service until January 1961. Photograph by Bill Reed.

LYME REGIS STATION – NO. 30583
In October 1958 Adams 415 Class no. 30583 has just reached Lyme Regis with a branch train from Axminster. Photograph from the Dave Cobbe Collection courtesy Rail Photoprints.

Above **MASBURY – NO. 53809 AND NO. 73049**
The 09.08 Bournemouth to Birmingham runs up towards Masbury summit with 7F Class no. 53809 and Standard Class 5 no. 73049 on 1st September 1962. Photograph by Hugh Ballantyne courtesy Rail Photoprints.

Below **MINEHEAD STATION – NO. 5522**
Collett 4575 Class no. 5522 stands at the head of a Taunton-bound train waiting for departure from Minehead in the late 1950s. Photograph by Bill Reed.

Above **MIDSOMER NORTON SOUTH STATION – NO. 53808**
Around halfway on the S&DJR's Bath Extension, Midsomer Norton was ready for traffic in July 1874. This was a year after the town was first served by the Bristol & North Somerset Railway's station (originally named Welton) on the line from Bristol to Radstock. Midsomer Norton station's name changed at the turn of the century to Midsomer Norton & Welton, as did the B&NSR in the early 1900s. At Nationalisation, this was deemed unhelpful for passengers and the S&DJR's changed to Midsomer Norton South. Fowler 7F Class no. 53808 has a Nottingham to Bournemouth (Saturday-only) train at the station on 4th August 1962. The engine was employed on the S&DJR from July 1925 to February 1964. Midsomer Norton South station was bought following closure in 1966 and managed to survive to the 1990s when restored by the Somerset & Dorset Railway Heritage Trust. Operating a short section of line, steam and diesel locomotives are used to haul trains for the public. Photograph by Dave Cobbe courtesy Rail Photoprints.

Below **NEWQUAY STATION – NO. 3635**

The Cornwall Minerals Railway developed a network of lines in the 1870s, stretching westward from Par to Newquay. As the name suggests, this was originally for mineral extraction, yet market volatility at the time saw the company struggle and the GWR was soon approached for help. The lines were initially leased before formal takeover in the late 1890s. The North West of Cornwall was untouched by the railways at this point but the GWR soon had a project to link several places from Chacewater on the main line to Perranporth and Newquay. The route was ready for traffic in 1905. Collett 8750 Class no. 3635 has used the latter to reach Newquay with a local passenger train from Chacewater a short time before closure in 1963. No. 3635 had been in Cornwall from 1949 though left for South Wales in 1962. Withdrawal from Tyseley occurred in May 1965 after a year there. Photograph by Bill Reed.

NEWTON ABBOT STATION – NO. 5920

An express hauled by no. 5920 *Wycliffe Hall* pauses for passengers at Newton Abbot station around 1960. Photograph by Bill Reed.

Above NEWTON ABBOT STATION – NO. 7029
No. 7029 *Clun Castle* has the 'Cornishman' express at Newton Abbot station, c. 1960. The engine is about to be coupled to a pilot, likely to tackle Dainton bank to the west. Photograph by Bill Reed.

Below NEWTON ABBOT STATION – NO. 5153
The pilot for no. 7029 *Clun Castle* is Collett 5101 Class 2-6-2T no. 5153. Apart from a sojourn in Slough, no. 5153 worked at Newton Abbot from the late 1940s to September 1962. Photograph by Bill Reed.

Opposite above **PAR STATION – NO. 5538**
A Newquay branch train is in the bay at Par station, c. 1960. Par station had been open on the Cornwall Railway in 1859, though the branch was not ready until 20 years later. Collett 4575 Class no. 5538 is at the head of the service. The locomotive was employed at Truro from January 1960 to June 1961. Despite two moves subsequently, the engine was condemned in October 1961. Photograph by Bill Reed.

Opposite below **PAR STATION – NO. 5028**
In the late 1950s, 'Castle' Class no. 5028 *Llantilio Castle* has an express at Par station. The engine had previously worked the 'Cornishman' as the headboard is stored on the running plate to the bottom right of the smokebox. In January 1959, no. 5028 was paired with a Hawksworth flat-sided tender and used until condemned in May 1960. Photograph by Bill Reed.

Below **NEWTON ABBOT STATION – NO. 1000**
Originally known as Newton, the station was opened by the South Devon Railway in late 1846 as the section from Teignmouth was ready to start operations. Around six months later construction had taken the railway westward to Totnes and at the end of the decade Plymouth was reached. The station's name changed to Newton Abbot in the late 1870s and by this time a branch from Moretonhampstead was connected just to the north. Hawksworth 'County' Class no. 1000 *County of Middlesex* has a main line express, c. 1960. Photograph by Bill Reed.

Above PAR STATION – NO. 6808

In the early 20th century Churchward's 4300 Class 2-6-0 was developed for GWR mixed traffic duties. By the 1930s traffic conditions were more strenuous and required an improved engine. Collett used the 'Hall' as a template and created the 'Grange' Class 4-6-0 with slightly smaller driving wheels and reconditioned parts from 4300 Class locomotives that were withdrawn to make way for the 'Granges'. A total of 80 class members appeared from Swindon from 1936 to 1939. No. 6808 *Beenham Grange* was sent into traffic during September 1936 and was employed to August 1964. The engine has a local train at Par station in the 1950s. A long-term employee at Penzance shed, no. 6808 was in South Wales from 1962 and had a spell in the Midlands before the end. Photograph by Bill Reed.

Opposite above PAR STATION – NO. 4585

A possible breakdown train is at Par station in the 1950s. The locomotive is Collett 4575 Class no. 4585. Built at Swindon in March 1927, the engine was in traffic over 30 years. In 1952, no. 4585 transferred to St Blazey and was there to October 1959 when sent for scrap. Photograph by Bill Reed.

Opposite below PENSFORD – NO. 3032

The Bristol & North Somerset Railway crossed the Chew Valley by a stone and brick viaduct. This was nearly 1,000 feet long, 100 ft tall, with 16 spans of irregular width. Opened in 1874, Pensford viaduct was in use to the late 1960s and has subsequently acquired listed status. A goods train for Radstock is on the viaduct during mid-March 1955 with GWR 3000 Class 2-8-0 no. 3032. The locomotive was one of 100 ex-Railway Operating Division locomotives purchased following the First World War. These had been built to help the war effort at home and abroad to J.G. Robinson's 8K Class design for the Great Central Railway. With a large number surplus, deals were made to clear the various dumps across the country. No. 3032 was bought in 1926. The engine was approaching the end here as withdrawal occurred in October. Photograph by Hugh Ballantyne courtesy Rail Photoprints.

Above PENRYN STATION – NO. 8421
A local train from Truro to Falmouth makes a stop at Penryn station with Hawksworth 9400 Class no. 8421 during June 1956. The locomotive was Exeter-allocated from new in 1950 to mid-1959 and the last six months were passed at Gloucester. No. 8421 had a disappointingly short service life of less than ten years. Photograph courtesy Rail-Online.

Below PENZANCE SHED – NO. 6849
At the head of a line of locomotives being serviced at Penzance shed during July 1956 is Collett 'Grange' Class no. 6849 *Walton Grange*. The engine was visiting from Truro. Photograph by Bill Reed.

Above PENZANCE STATION – NO. 6809

View from Chyandour Cliff to Penzance station in July 1956. Collett 'Grange' Class no. 6809 *Burghclere Grange* stands at the head of an express before departure. The 'Grange' Class was built in two lots, with no. 6809 part of the first for 60 engines built in the mid-1930s. The second comprised 20 erected at the end of the decade. No. 6809 was produced at Swindon in September 1936. During early 1950, the locomotive began the first of eight years at Penzance depot. After leaving, the engine had spells at Bristol and South London. Photograph by Bill Reed.

Above PENZANCE STATION – NO. 5021

Only a single platform was provided at Penzance station for the opening of the West Cornwall Railway between Redruth and Penzance in March 1852. In a departure for the area, the company used standard gauge, though broad gauge was later added in tandem and persisted until the 1890s. When the GWR took over, the company completely remodelled the station and erected a two-platform structure using granite, with the plans created by company architect W.L. Owen. In the 1930s, the station was further expanded and another two platforms were installed. 'Castle' Class no. 5021 *Whittington Castle* is coupled to an express awaiting departure from Penzance in the late 1950s. An engineman is in the Collett tender, though from late 1958 a Hawksworth type was attached and used to withdrawal in September 1962. The locomotive was likely Plymouth-allocated at the time of the image though was soon to move to Cardiff. Photograph by Bill Reed.

Opposite PENZANCE SHED – NO. 6873

When the West Cornwall Railway was completed in the early 1850s, an engine shed was built by the company next to the station. A larger building was necessary a decade later and this was erected a short distance to the east. After the GWR took over the company in 1876, a new shed was completed on the same site and lasted until 1914 when another depot was required. Land around a mile to the east of the station was purchased and used for a four-road building with associated facilities. Steam lasted until September 1962, though further use for diesel locomotives saw the shed open until 1976. The site has been redeveloped for retail use. Collett 'Grange' no. 6873 *Caradoc Grange* is outside the shed around 1960 when based at Plymouth. The locomotive had been at Penzance for around a month in the winter of 1958/1959. Photograph by Bill Reed.

PENZANCE STATION – NO. 6855

The 'Cornishman' stands against the platform at Penzance station in the late 1950s with 'Grange' Class no. 6855 *Saighton Grange*. Photograph by Bill Reed.

Above **PENZANCE SHED – NO. 6808**
At Nationalisation, around six 'Granges' were based at Penzance, though this number had doubled by the end of the 1950s. One of two in the yard, c. 1960, is no. 6808 *Beenham Grange*. Photograph by Bill Reed.

Below **PENZANCE SHED – NO. 8473**
Seemingly abandoned at Penzance shed is 9400 Class no. 8473. The engine was less than ten years old when condemned in January 1961. Initially sent to Penzance, no. 8473 was soon moved on to Newton Abbot. Photograph by Bill Reed.

Above **PLYMOUTH – NO. 1361**
North west of Plymouth station, Churchward 1361 Class 0-6-0ST no. 1361 is shunting wagons at Keyham station during June 1959. Photograph courtesy Rail-Online.

Below **PLYMOUTH STATION – NO. 5034**
On 4th August 1958, an engineman checks the motion of no. 5034 *Corfe Castle* at Plymouth. Photograph by Bill Reed.

Above **PLYMOUTH STATION – NO. 5532**
Light engine at the east end of Plymouth station on 2nd September 1960 is 4575 Class no. 5532. Photograph by Bill Reed.

Below **PORTISHEAD STATION – NO. 4535**
A local to Bristol is ready to leave Portishead station with 4500 Class no. 4535 in November 1952. Portishead power station is in the background and expansion in the mid-1950s caused the railway station to be moved westward. Photograph courtesy Rail Photoprints.

Above **RADSTOCK WEST STATION – NO. 4636**
Radstock was around halfway between Bristol and Frome and the point where the lines between the two met following construction by independent companies. The S&DJR also crossed the route on the way from Evercreech to Bath. Radstock station on the Bristol & North Somerset Railway was opened during September 1873, whilst the S&DJR facility accepted travellers during the following year. At Nationalisation the names of the two had to be changed and became Radstock West and Radstock North respectively. Passenger trains were withdrawn from Radstock West in 1959, but freight continued until early 1966. Collett 8750 Class no. 4636 takes water at Radstock West after working the 07.30 freight from Frome on the last day of steam operations on the line – 4th September 1965. The locomotive survived for just a few more days. Photograph by Derek Fear courtesy Rail Photoprints.

Opposite **PLYMOUTH STATION – NO. 4591**
Plymouth was first served by a station on the South Devon Railway. This operated from April 1849 to 1876. At this time, the SDR was taken over by the GWR and the L&SWR arrived leading to the construction of a joint station. The latter company later opened Plymouth Friary as a terminus and this was in use to 1958 when services concentrated at the joint station. Collett 4575 Class no. 4591 is seen at Plymouth c. 1960. The locomotive was built in March 1927 and served to August 1964. No. 4591 worked locally until March 1963 when transferred to Salisbury. Photograph by Bill Reed.

Above **RADSTOCK NORTH STATION – NO. 80013**
With the closure of Radstock West in 1959, passengers had to rely on Radstock North until 1966 when the whole S&DJR was closed. On 1st July 1965, BR Standard Class 4 2-6-4T no. 80013 is leaving Radstock North with the 09.05 Templecombe to Bath Green Park passenger train. No. 80013 arrived at Bournemouth from Brighton in June 1964 and was dispatched from there for scrapping during September 1966. Photograph by Derek Fear courtesy Rail Photoprints.

Opposite page **SALISBURY – NO. 34023**
Gresley A4 Class Pacific no. 4498 *Sir Nigel Gresley* left service in February 1966. The locomotive was purchased by the A4 Locomotive Society and refurbished at Crewe Works. Returning to steam in April 1967, no. 4498 has been a mainstay of the preservation since. On 3rd June 1967 no. 4498 began the first of two days touring the Southern Region. Leaving Waterloo, the engine worked to Southampton where Bulleid 'West Country' Pacific no. 34023 *Blackmore Vale* was used for the short journey to Salisbury, at which point no. 4498 again took over. A similar route was travelled for the railtour on the following day. No. 34023 was condemned in July 1967 and also preserved, being bought by the Bulleid Preservation Society. In steam for two periods over the ensuing years, no. 34023 has been stopped since 2008 with firebox problems, yet with a replacement recently made, hopefully the engine will be back in steam within the next few years. Both photographs by Geoff Warnes.

Above **SALISBURY STATION – NO. 34100**
On 26th June 1966, 'West Country' Class no. 34100 *Appledore* was one of three Bulleid Pacifics used on the 'Devonshire Rambler' railtour. No. 34002 *Salisbury* started the day off at Waterloo and transported the party to Salisbury. 'Merchant Navy' Class no. 35023 *Holland-Afrika Line* then travelled westward to Exeter on the L&SWR line, then returned eastward via Taunton to Westbury. No. 34100 was waiting and took the outing back to Waterloo via Salisbury and Southampton. The locomotive has made a stop at Salisbury to take water. The 4,500-gallon capacity is breached as the fireman looks on while rearranging coal in the space which could hold five tons. The tender would have been streamlined with the engine, but this was removed when rebuilding took place in September 1960. Photograph by P.C. Wakefield courtesy Colour-Rail.

Opposite above **SALISBURY STATION – NO. 7922**
The L&SWR was the first to reach Salisbury as a connection was laid from Eastleigh in 1847. Nine years later, the GWR reached the west of Salisbury after the construction of a route from Westbury. The L&SWR opened a second line to Salisbury at the end of the decade to reduce the distance to the city and this forged southward from Andover. As this was completed, a new station opened and sited further to the west of the original next to the GWR station to allow improved exchange of goods. Not until the 1930s did the GWR start using the L&SWR station which had been expanded several times in the ensuing years. Hawksworth 'Modified Hall' no. 7922 *Salford Hall* is light engine at Salisbury station in May 1963. The locomotive was based at Southall from late 1962 until 1965 and had a spell at Oxford before condemned at the end of the year. Photograph courtesy Rail Photoprints.

Opposite below **SALISBURY STATION – NO. 76011**
BR Standard Class 4MT 2-6-0 no. 76011 has been attached to the Saturday-only 09.35 Swansea to Brockenhurst train on 7th July 1963. The service left Salisbury and ran southward via Wimborne and Bournemouth. No. 76011 was Eastleigh-allocated at this time and later transferred to Bournemouth in 1965. The engine was condemned there in July 1967. Photograph by B.W.L. Brooksbank.

SALISBURY SHED –
NO. 30064 AND NO. 30072
Two S100 Class 0-6-0Ts are in the yard at Salisbury shed. No. 30064 and no. 30072 were both condemned in July 1967. Photograph by Bill Reed.

Above SALISBURY STATION – NO. 34054
Two extra pairs of hands have been enlisted to make sure 'Battle of Britain' Pacific no. 34054 *Lord Beaverbrook* is ready to continue on 29th August 1964. The locomotive had around a week left in traffic. Photograph by David Christie.

Below SALISBURY STATION – NO. 3864
Passing Salisbury signal box light engine on 29th August 1964 is Churchward 2800 Class no. 3864. Built in November 1942, the locomotive was in service to July 1965. Photograph by David Christie.

Above SALISBURY STATION – NO. 34038
The East Midlands branch of the Railway Correspondence & Travel Society arranged for members to journey to the South Coast on 9th May 1964. The 'East Midlander' left Nottingham Victoria with Stanier 'Coronation' Pacific no. 46251 *City of Nottingham* and used the ex-Great Central main line and branch to Banbury, then taking the line to Oxford and Didcot. At the latter, 'West Country' Pacific no. 34038 *Lynton* ran southward via Newbury and Winchester to Eastleigh for a tour of the works and shed. The engine has reached Salisbury here as the tour headed for Swindon. There, no. 46251 returned to the train for the journey back to Nottingham. Photograph by Bill Reed.

Opposite above SALISBURY STATION – NO. 76065 AND NO. 76064
Two BR Standard Class 4 2-6-0s are at Salisbury station on 27th July 1963. The design was developed from H.G. Ivatt's Class 4MT 2-6-0 built for the London Midland & Scottish Railway. A total of 115 locomotives appeared from three workshops between 1952 and 1957. No. 76065, on the right, was erected at Doncaster Works in July 1956, as was classmate no. 76064, to the left. The pair were new to Eastleigh shed. No. 76065 has picked up the 09.32 Cardiff to Pokesdown Saturday-only train and will travel via Bournemouth, whilst no. 76064 stands at the head of a service to Waterloo. Photograph by B.W.L. Brooksbank.

Opposite below SALTASH STATION – NO. 6406
Collett 6400 Class no. 6406 has a local train at Saltash station, west of Plymouth, on 4th August 1958. Introduced in 1932, the 6400 Class numbered 40 examples all built during the year. The main feature of these engines was being equipped to work auto-trains. A long-term Plymouth resident, no. 6406 was condemned in June 1960. Photograph by Bill Reed.

Above **SAVERNAKE LOW LEVEL STATION – NO. 9740**
Promoted as a strategic line to compete against the L&SWR, the Berks & Hants Railway was backed by the GWR and ran from Reading to Devizes, east of Bath. Savernake station opened with the line during November 1862. Two years later, a branch to nearby Marlborough was built and Savernake acted as a junction with several trains running daily. This is the service Collett 8750 Class no. 9740 is waiting with in the bay platform on 8th August 1961. In the late 19th century, the Midland & South Western Junction was laid to connect the two companies and a second station was built at Savernake. After Grouping, the GWR station became Low Level and later in 1961 Savernake for Marlborough when High Level closed. This fate was five years away for the 1862 station. Photograph by Bill Reed.

Opposite above **SAVERNAKE LOW LEVEL STATION – NO. 6327**
The Churchward 4300 Class 2-6-0 design was formulated from several classes introduced around the turn of the century. A powerful boiler was fitted to an existing chassis with standard wheels and the type was intended for use on mixed duties. The first class member appeared in 1911 and a steady stream emerged from Swindon Works up to Grouping, with two batches completed afterwards. No. 6327 was part of a group of 24 locomotives built at Swindon to lot no. 212 in 1921. The engine was approaching the end on 8th August 1961, with two years left in steam. No. 6327 approaches Savernake Low Level station with a local freight train. At this time, the locomotive was employed at Swindon, though spent the final year in Taunton. Photograph by Bill Reed.

Opposite below **SEATON SHED – NO. 30021**
A small engine shed was provided at the end of the Seaton branch, which ran from the main line west of Axminster. The Seaton & Beer Railway opened this in 1868 on a site just to the north east of the station. In the mid-1930s, the Southern Railway replaced the timber structure with one made from concrete blocks (closer to the station) and this is visible in the background. Drummond M7 Class no. 30021 stands at the coal stage, with large blocks of fuel present, during April 1958. The engine had a long-term allocation to Exmouth Junction which lasted to January 1963. Moving to Tunbridge Wells, six months were spent there before a transfer to Salisbury occurred. No. 30021 was withdrawn there during March 1964. Photograph by Bill Reed.

Below SHEPTON MALLET – NO. 92001

Intended for heavy freight duties between marshalling centres, BR's Standard Class 9F 2-10-0 design found a secondary role as a passenger locomotive during holiday seasons. Early in their career, 9Fs were called on to assist with failures and perhaps inspired by this, the operating department began scheduling 9Fs in the 1957 summer season for excursions. Previously, these trains had been left to whatever motive power was available and in places the locomotives had been found in deficit of power. The 9Fs were initially popular on the Midland Main Line and in 1960 experiments were made on the S&DJR with exchange services from Bristol to Bournemouth. No. 92001 joined the ranks at Bath in 1961 and 1962 just for the traffic, working elsewhere in the winter. On 1st September 1962, the locomotive passes Shepton Mallet with an express from Bradford to Bournemouth. Photograph by Dave Cobbe courtesy Rail Photoprints.

Above SHARPNESS STATION – NO. 1445

To cut out a bend in the River Severn, the Gloucester & Sharpness Canal Company was formed in the late 18th century. Eventually completed in 1827, access from the river occurred at Sharpness which was not historically associated with docks or port activities. Despite the Bristol & Gloucester Railway passing just four miles to the east, a branch was not built to Sharpness until 1875. This later became a through route after the completion of the Severn Bridge in 1879. Sharpness station was originally opened in 1876 as a terminus. Rebuilding had to take place for through traffic to commence. Collett 1400 Class no. 1445 has a local train at Sharpness on 15th July 1964. Passenger services were withdrawn on 31st October 1964, whilst freight lingered to 1966. The line has been subject to plans for a heritage railway in the early 2010s and a new station at Sharpness to serve a new housing development. Photograph by R.C. Riley courtesy Rail-Online.

Above SHAPWICK STATION – NO. 41296
The 13.15 from Evercreech Junction to Highbridge approaches Shapwick station on 11th December 1965. Ready to exchange tokens, the engineman is on Ivatt Class 2MT no. 41296. The locomotive had been on the S&DJR from 1957. Photograph by Hugh Ballantyne courtesy Rail Photoprints.

Below SIDMOUTH STATION – NO. 41307
A branch train is at Sidmouth station with Ivatt Class 2MT no. 41307. The locomotive was at Exmouth Junction shed for ten years, 1955-1965, then witnessing the end of the S&DJR at Templecombe. Photograph by Bill Reed.

Above **SIDMOUTH STATION – NO. 82019**
Authorised in 1862, the Sidmouth Railway encountered financial difficulties and failed at the end of the decade. Under fresh management, the line was able to reach completion in 1874. No. 82019 has a local train at Sidmouth, c. 1960. Photograph by Bill Reed.

Below **SIDMOUTH JUNCTION STATION – NO. 30796**
Feniton was the original name for Sidmouth Junction station and several variations were used before the opening of the branch in 1874. Maunsell 'King Arthur' no. 30796 *Sir Dodinas Le Savage* passes with an Exeter to Salisbury local, c. 1960. Photograph by Bill Reed.

Above ST ERTH STATION – NO. 4563
The nearest point of the West Cornwall Railway to St Ives was around four miles away and the station established to serve the town was called St Ives Road. In 1877 a branch was finally opened and the main line station was renamed St Erth. Churchward 4500 Class no. 4563 arrives at St Erth with a train during July 1956. The engine was allocated to Penzance, arriving there in December 1954 and remaining employed until condemned during October 1961. Photograph by Bill Reed.

Opposite above SIDMOUTH JUNCTION STATION – NO. 30024
Nearly 20 years elapsed from the opening of the Sidmouth Railway to the start of a project to extend westward along the coast to Budleigh Salterton. Unlike the other line, this short branch from Tipton was completed swiftly for opening in 1897. Six years later, the L&SWR built a short line to connect Budleigh Salterton with Exmouth. Drummond M7 Class no. 30024 is bound for Exmouth here, leaving Sidmouth Junction station on 31st July 1960. Photograph by Dave Cobbe courtesy Rail Photoprints.

Opposite below ST ERTH STATION – NO. 1028
Hawksworth's 'County' Class 4-6-0 featured a particularly high boiler pressure – 280 lb per sq. in. – when new but this was later reduced to 250 lb per sq. in., yet still remained above normal for a GWR design. The class members also had a double chimney. Just 30 were built between 1945 and 1947. No. 1028 *County of Warwick* was one of the last in traffic during March 1947. The locomotive has the 'Cornishman' at St Erth station in the late 1950s. Bristol Bath Road-allocated, no. 1028 moved across to St Philip's Marsh when the first mentioned closed to steam in September 1960. Photograph by Bill Reed.

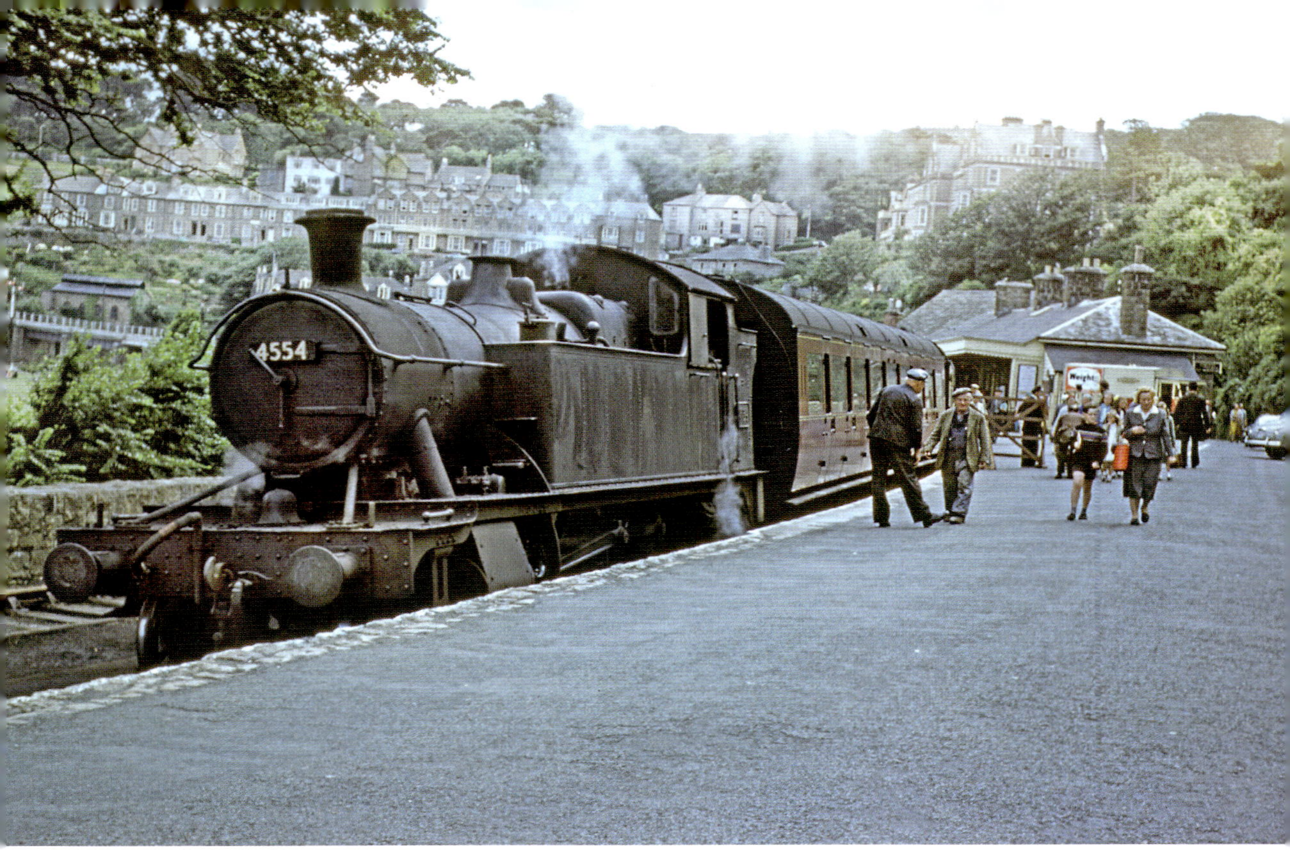
4554

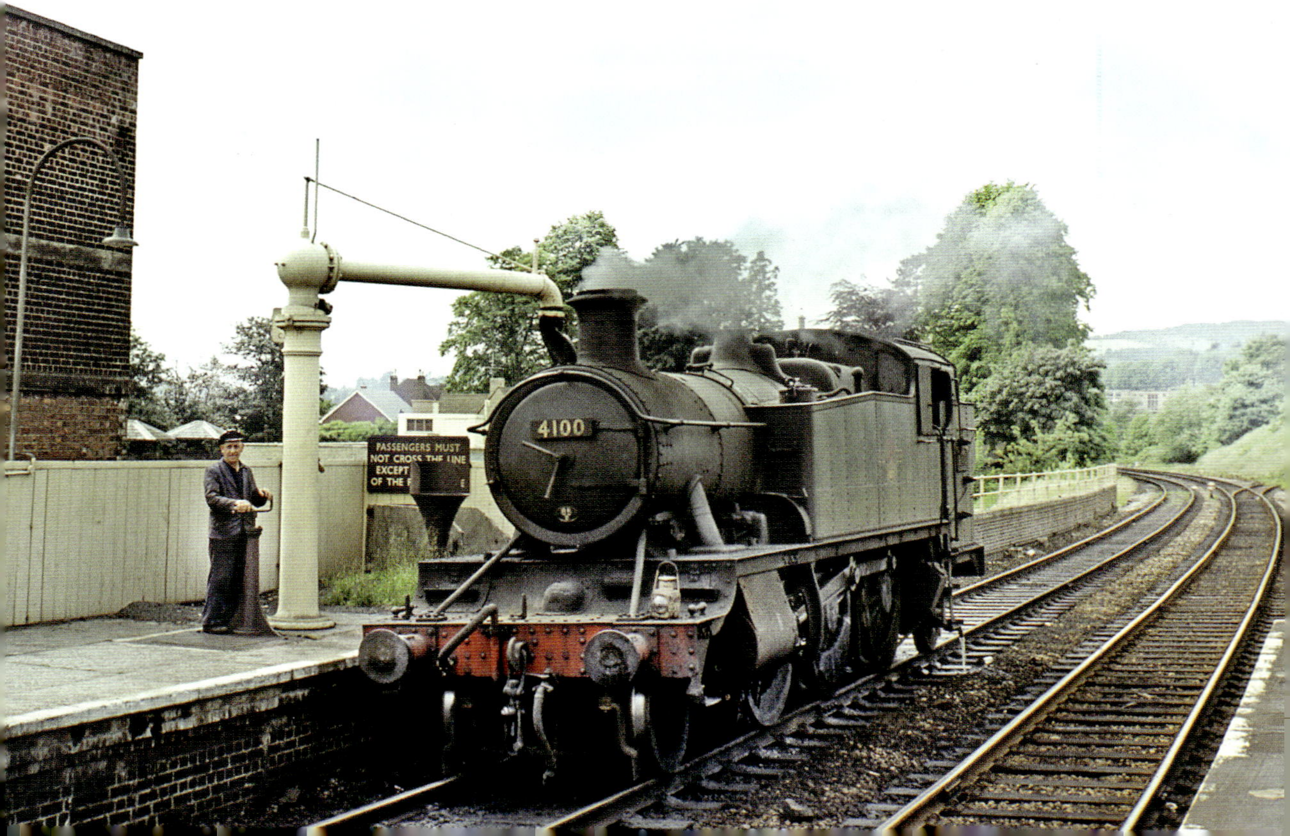
4100
PASSENGERS MUST
EXCEPT
OF THE

Above **SWANAGE STATION – NO. 30106**
Many Drummond M7 Class locomotives ended their careers working the Swanage branch. No. 30106 was no exception and was condemned in November 1960. This event was a year away as the engine has a local train at Swanage on 1st September 1959. Two class members have been preserved subsequently, with one employed on the Swanage branch which operates as a heritage railway. Photograph by Geoff Warnes.

Opposite above **ST IVES STATION – NO. 4554**
Known for the fishing industry in the 19th century, St Ives subsequently developed as a tourist destination and welcomes over 500,000 people annually at present. This has helped to keep the branch line operational whereas many in the area have been closed. Though the route was originally planned in the 1840s, not until 1877 did the scheme reach completion as the last new broad gauge project. The line was threatened by Dr Beeching but managed to evade the axe. The station was rebuilt in the early 1970s and now has a straight platform rather than curved as seen in this image. Churchward 4500 Class no. 4554 has a local train in the late 1950s. Photograph by Bill Reed.

Opposite below **STROUD STATION – NO. 4100**
In the late 1920s, Collett introduced his local passenger design – the 5100 Class 2-6-2T. Production stretched just past Nationalisation, though a pause occurred during the Second World War. In total 140 locomotives were produced. No. 4100 was the first of a batch of ten built in 1935. The engine is between duties here and is taking water at Stroud station on 26th July 1964. No. 4100 was at Gloucester Horton Road from July 1957 and was condemned there in October 1965. Photograph courtesy Colour-Rail.

Above SWINDON STATION – NO. 2838

The Pennsylvania Railroad was the first to use the 2-8-0 wheel arrangement in the mid-1860s, though this was unusual in having a rigid axle at the front. The adoption of the pony truck improved the design as this reduced stresses at the front end. Eight coupled wheels also suited heavy freight trains as power could be transferred more efficiently when starting off. Not until the early 20th century did Britain see a 2-8-0 and this was pioneered by G.J. Churchward with the 2800 Class. A prototype appeared in 1903, with a production series built from 1905-1919. At the latter date, the total in service was 84 engines. No. 2838 was completed at Swindon Works in September 1912 and was in steam until August 1959. An unfitted freight runs through Swindon station behind no. 2838 in the late 1950s. Photograph by Bill Reed.

Opposite above SWANAGE STATION – NO. 30105

The Southampton & Dorchester Railway opened in the 1840s and declined to build a branch to Swanage. The industrial businesses in the area were encouraged to open a line independently though the scheme failed. In the early 1880s, a route from Wareham to Swanage received an Act of Parliament and construction was completed swiftly for opening in 1885. Passenger and freight traffic was developed subsequently, with summer tourist services and clay extraction dominating. Drummond M7 Class no. 30105 has a local at Swanage station in the late 1950s. The line managed to survive the Beeching Axe but BR was persistent and the final passenger train operated on 1st January 1972. Soon after, preservation attempts began and the Swanage Railway Society has been successful in re-establishing the route throughout. Photograph by Bill Reed.

Opposite below SWINDON STATION – NO. 5963

A parcels train is at Swindon station, c. 1960, with 'Hall' Class no. 5963 *Wimpole Hall*. On the smokebox door is Westbury's '82D' shed code and this was present from February 1956 to February 1963. The locomotive was then at Bristol St Philip's Marsh and this lasted to June 1964 when sent for scrap. Photograph by Bill Reed.

Above SWINDON WORKS – NO. 3210
No. 3210 was amongst the final lot of 2251 Class 0-6-0s which numbered 20. The locomotive was built at Swindon in December 1947 and is under attention in the works here. No. 3210's service life approached nearly 17 years. Photograph by Bill Reed.

Opposite above SWINDON STATION – NO. 5715
Collett 5700 Class no. 5715 is seen at Swindon station on 17th August 1954. Constructed by the North British Locomotive Company in March 1929, the engine was in traffic to August 1958. Photograph by Bill Reed.

Opposite below SWINDON STATION – NO. 2211
Passing through Swindon station in the late 1950s is 2251 Class no. 2211. Appearing from Swindon Works in May 1940, no. 2211 survived to November 1964. At the time of the image, the locomotive was likely employed at Exeter. Photograph by Bill Reed.

Above SWINDON STATION – NO. 73036
BR Standard Class 5 no. 73036 appears to have a freight service running through Swindon station. New to Carlisle in September 1953, the locomotive transferred to the Western Region a month later and remained until condemned during September 1965. Photograph by Bill Reed.

Below SWINDON STATION – NO. 6103
Collett 6100 Class no. 6103 is outside the Erecting Shop at Swindon Works on 4th September 1959. Just in view to the right is the cab of 4300 Class no. 5385. Photograph by Geoff Warnes.

Above SWINDON WORKS – NO. 7029
No. 7029 *Clun Castle* has completed a repair at Swindon, c. 1960. Withdrawn in late 1965, the engine was preserved the following year and is currently operational. Photograph by Bill Reed.

Below SWINDON WORKS – NO. 6018
Out of service at Swindon Works is 'King' Class no. 6018 *King Henry VI*. Withdrawn in late 1962, the engine is seen during June 1963 and disposal occurred in October. Photograph by Geoff Warnes.

Above SWINDON WORKS – NO. 6331
The main railway workshops generally disposed of their own stock when the time came. Swindon had a dedicated shop which dealt with scrapping locomotives, though the volume briefly overwhelmed the team and the works pioneered the outsourcing of the task. In 1959, a batch was sent to Woodham Bros Scrapyard, Barry, South Wales and many ex-GWR followed to the end of steam. On 4th September of the year, 4300 Class no. 6331 is in the yard waiting for the torch. Withdrawal from Banbury had taken place in April with 38 years to the engine's credit. Ex-LMSR and ex-LNER locomotives were also scrapped at Swindon, though the activity stopped in mid-1965. Photograph by Geoff Warnes.

Opposite above TAUNTON STATION – NO. 2277
The Bristol & Exeter Railway opened as far as Taunton on 1st July 1842 and the town boasted a station designed by Brunel. Following the opening of several local lines, the facility had to be extensively rebuilt in the late 1860s, whilst more changes were made in the 1890s. During the early 1930s, the number of running lines either end of Taunton station was doubled leading to the construction of a central island platform and new north-facing platform with station buildings. A passenger train is at the island platform on 30th August 1960 with 2251 Class no. 2277 at the head. The engine was based locally from June to July 1961. Photograph by Bill Reed.

Opposite below TAUNTON STATION – NO. 4604
Also working at Taunton on 30th August 1960 is 8750 Class no. 4604. The locomotive was new in October 1941. Just after Nationalisation, no. 4604 was received at Taunton and remained to August 1962. The engine left traffic after three years in South Wales. Photograph by Bill Reed.

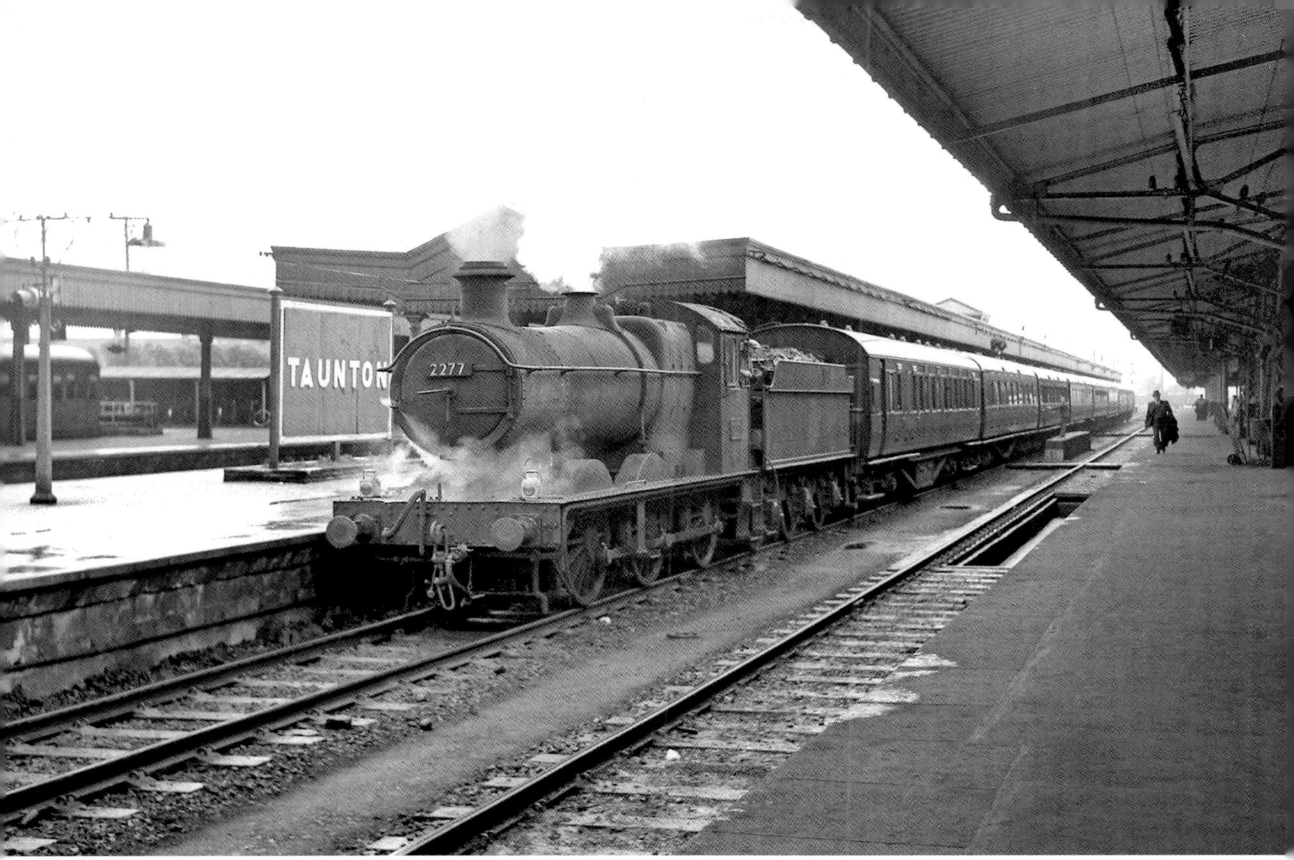
TAUNTON
2277

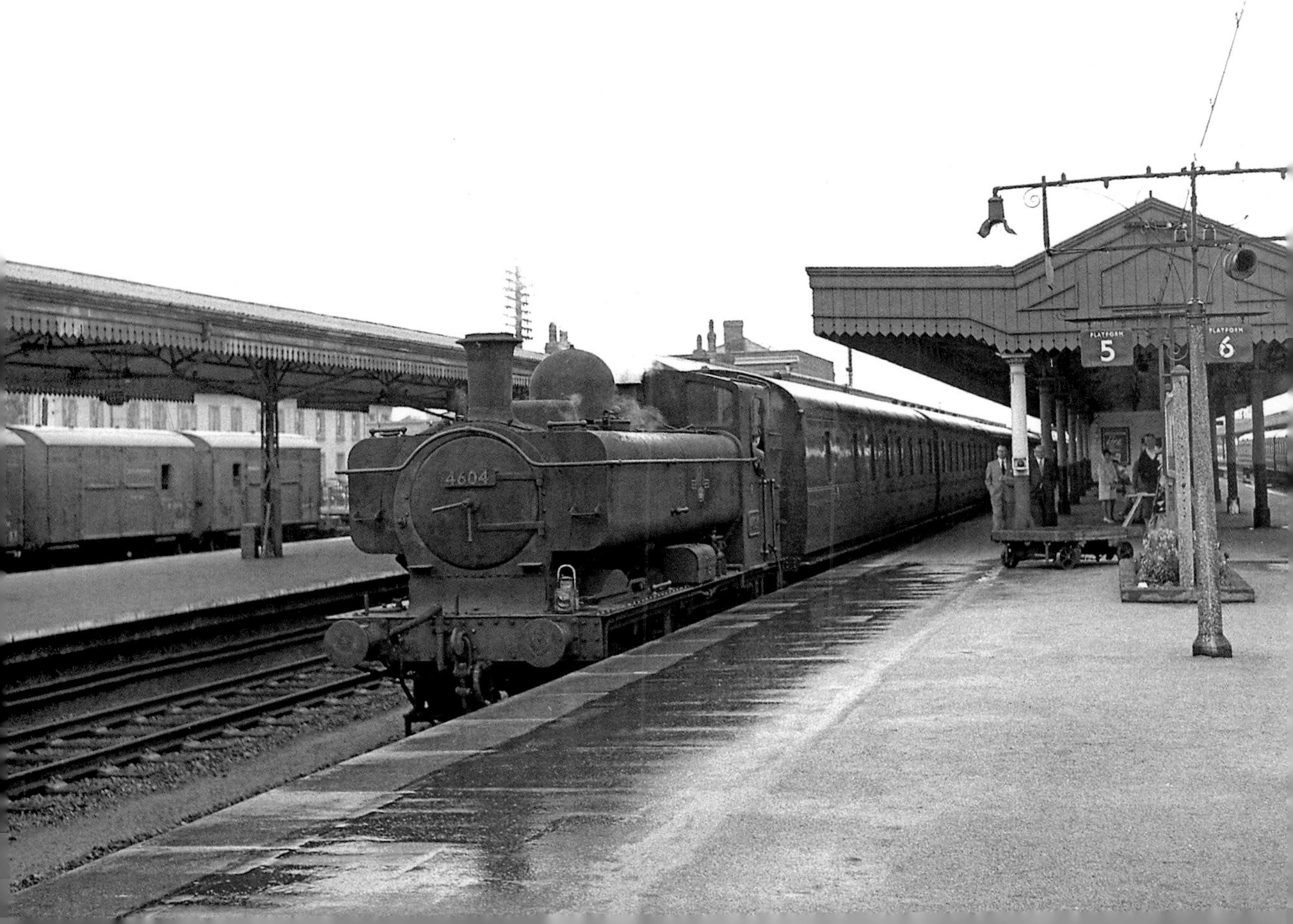
4604
PLATFORM
5
PLATFORM
6

TAVISTOCK NORTH STATION NO. 34011

On 17th July 1962, 'West Country' Pacific no. 34011 *Tavistock* is at Tavistock North station. Photograph courtesy Colour-Rail.

Above **TAVISTOCK SOUTH STATION – NO. 5572**
A local train is at the GWR's Tavistock South station with 4575 Class no. 5572. The facility was closed at the end of 1962, whilst North survived to the late 1960s. Photograph by Bill Reed.

Below **TEMPLECOMBE – NO. 82002**
Passing the engine shed, no. 82002 arrives at Templecombe with a local from Bournemouth on 4th August 1962. Photograph from the Dave Cobbe Collection courtesy Rail Photoprints.

Above **TEMPLECOMBE STATION – NO. 73051**
A Bournemouth train stands against the platform at Templecombe station on 3rd July 1965, with Standard Class 5 no. 73051 at the head. Photograph by David Christie.

Below **TEMPLECOMBE STATION – NO. 80041**
Standard Class 4MT no. 80041 arrives at Templecombe with empty stock for a working to Bath on 3rd July 1965. Photograph by David Christie.

Above **TEWKESBURY STATION – NO. 7756**
Enginemen take a break at Tewkesbury station before the service to Ashchurch leaves on 25th March 1961. The locomotive is 5700 Class no. 7756. Photograph by B.W.L. Brooksbank.

Below **TAVY BRIDGE – NO. 34104**
Just north of Plymouth, the Tavy bridge crossed the Tavy estuary. 'West Country' Pacific no. 34104 *Bere Alston* is on the structure with the 09.00 service from Waterloo to Plymouth on 4th August 1960. Photograph by Hugh Ballantyne courtesy Rail Photoprints.

TIVERTON JUNCTION STATION – NO. 1440

A Tiverton branch train is at Tiverton Junction station with 1400 Class no. 1440. Photograph by Bill Reed.

Above **TIPTON ST JOHNS STATION – NO. 82011**
A train from Exmouth has reached Tipton St Johns station with BR Standard Class 3 no. 82011. Photograph by Bill Reed.

Below **TIVERTON JUNCTION STATION – NO. 1434**
A mixed Hemyock branch train stands against the platform at Tiverton Junction station. 1400 Class no. 1434 is the engine employed. Photograph by Bill Reed.

TRURO STATION – NO. 6873
A through freight is at Truro station with 'Grange' Class no. 6873 *Caradoc Grange*. Photograph by Bill Reed.

Above **TIPTON ST JOHNS STATION – NO. 82011**
A train from Exmouth has reached Tipton St Johns station with BR Standard Class 3 no. 82011. Photograph by Bill Reed.

Below **TIVERTON JUNCTION STATION – NO. 1434**
A mixed Hemyock branch train stands against the platform at Tiverton Junction station. 1400 Class no. 1434 is the engine employed. Photograph by Bill Reed.

Above TRURO STATION – NO. 5972

Following the successful introduction of the 'Castle' Class for express passenger services across the GWR system, Collett turned to a medium-powered mixed traffic design. The predecessor to this type was the 'Saint' Class 4-6-0 and in 1924 one was rebuilt with new specifications to prove them in traffic. When determined suitable an initial 80 were erected in the late 1920s. A number of orders were placed in the 1930s and into the early 1940s. When production ceased 258 'Hall' Class 4-6-0s were in service and a mainstay of services on the principal lines. No. 5972 *Olton Hall* was amongst a group of ten that appeared from Swindon in 1937, being ready in late April. The engine is at Truro station during July 1956 with an express. Allocated to Penzance, a transfer to Plymouth occurred in November 1958. By mid-1960, the locomotive had reached South Wales and was at several locations to the end of 1963. In the early 1980s, no. 5972 was rescued from the scrapyard and returned to steam just before the new millennium. The engine has since become the star of a popular children's film series. Photograph by Bill Reed.

Opposite above TORRINGTON STATION – NO. 41298

On 1st August 1854 the North Devon Railway opened between Crediton and Barnstaple. A line onwards to Bideford was part of the authorising Act, but the company passed on the project to the independent Bideford Extension Railway which first operated trains during November 1855. Another connection was made with a line from Bideford to Great Torrington in mid-1872. By this time the NDR had been absorbed by the L&SWR, forming part of the company's network. The station at Great Torrington was known as just Torrington and became an important point for milk traffic. Some of the tankers used are seen off to the right. These trains ran on the line to 1978 whilst other industrial locations kept trains moving until 1982. Passenger services had been discontinued in 1965. Ivatt Class 2MT no. 41298 has stopped at Torrington on 11th June 1963 with a freight service and is taking water. The Tarka Valley Railway has subsequently acquired the Torrington site with a view to operating a heritage line to Bideford. Photograph by L. Rowe courtesy Colour-Rail.

Opposite below TRURO STATION – NO. 6387

'Grange' Class no. 6387 *Forthampton Grange* has arrived at Truro station with a local train during July 1956. The engine was Penzance-allocated at this time, arriving there four years earlier from South Wales. After a few months in Plymouth, no. 6387 returned there in late 1960. In July 1965 the locomotive was sent for scrap. Photograph by Bill Reed.

TRURO STATION – NO. 6873
A through freight is at Truro station with 'Grange' Class no. 6873 *Caradoc Grange*. Photograph by Bill Reed.

Above **TRURO STATION – NO. 1007**
Truro-allocated 'County' Class no. 1007 *County of Brecknock* is at the station in the late 1950s. Photograph by Bill Reed.

Below **TRURO STATION – NO. 4549**
4500 Class no. 4549 takes water at Truro station on 22nd July 1960. Penzance-allocated at this time, the engine moved to Truro in September. Photograph by R.C. Riley courtesy Rail-Online.

Above **WADEBRIDGE – NO. 30586**
In between duties at Wadebridge, c. 1960 is Beattie 0298 Class 2-4-0WT no. 30586. The engine was built by Beyer Peacock & Co. in November 1875 and impressively lasted until December 1962. Photograph by Bill Reed.

Opposite above **WADEBRIDGE SHED – NO. 30708**
The Bodmin & Wadebridge Railway established the first servicing facilities for locomotives at Wadebridge in 1834 and these were used to the end of the 19th century. The L&SWR added a two-line depot to the east side of the station at this time and a decade later alterations were caried out. BR fully refurbished the shed just after Nationalisation and locomotives continued to be serviced there until late 1964. Drummond T9 Class 4-4-0 no. 30708 receives coal from a dispenser at Wadebridge shed during the mid-1950s. Withdrawal from Exmouth Junction occurred in December 1957. Photograph by Bill Reed.

Opposite below **WADEBRIDGE STATION – NO. 4666**
After nearly a decade in South Wales covering the 1950s, 8750 Class no. 4666 was in the West Country from November 1959. First at Exmouth, then at the end of the year Wadebridge took over the locomotive's allocation. No. 4666 worked there until February 1963 when returned to Exmouth and remained – apart from three months at Barnstaple – to June 1965 when sent for scrap. The locomotive is at Wadebridge station with a local on 1st September 1960. Photograph by Bill Reed.

Below WADEBRIDGE – NO. 30587

J.H. Beattie designed the 0298 Class 2-4-0WT for London suburban traffic in 1863. These were the result of several experiments to determine the best type for the duty. The 0298 Class was built in small batches consistently over 12 years when 85 locomotives had been completed. Most of these came from Beyer, Peacock & Co., though three appeared from Nine Elms Works in 1872. Despite being strong performers, the suburban services soon required new motive power. Some were converted to tender engines, whilst others were refurbished for use in the South West. A trio were taken to the Bodmin & Wadebridge Railway where the 0298 Class was found to be eminently suited to the route. No. 30587 was amongst the number and remained in steam to December 1962. The locomotive is at Wadebridge on 1st September 1960. Of interest between no. 30587 and no. 30586 (p. 141), is the different placement of the number and BR emblem, with the latter on the bunker here. No. 30587 was later preserved as part of the National Collection and has spent time on the Bodmin & Wenford heritage railway. Photograph by Bill Reed.

Above **WADEBRIDGE STATION – NO. 5502**
At the dawn of the railways, the Bodmin & Wadebridge Railway was founded to transport minerals from Bodmin and Wenfordbridge to the River Camel at Wadebridge. Opened in September 1834, a passenger service was also offered but just between Bodmin and Wadebridge and at intervals of days. The company was taken over in the following year, but the new ownership only lasted a decade before the L&SWR took possession in a strategic move. Not until the 1890s was the B&WR connected to the main line thanks to the North Cornwall Railway which built the line from Padstow to Halwill via Wadebridge. A station at the latter was opened in 1834 but was rebuilt in the late 1880s as part of the NCR's activities. During track improvements on the Bodmin line, the station was again altered. By the late 1880s, the GWR and L&SWR were able to form amicable relations and a connection was made between the two lines at Bodmin. In the mid-1950s, 4575 Class no. 5502 has likely used the junction to reach Wadebridge with a local train in the mid-1950s. The locomotive was allocated to St Blazey at the time. Photograph by Bill Reed.

WENFORDBRIDGE – NO. 30585
China clay wagons are with 0298 Class no. 30585 at Wenfordbridge, c. 1960. Photograph by Bill Reed.

Above WAREHAM STATION – NO. 30060
A Swanage branch train has arrived at Wareham station with Drummond M7 Class no. 30060 during April 1958. Photograph by Bill Reed.

Below WAREHAM STATION – NO. 35021
Rebuilt 'Merchant Navy' Class Pacific no. 35021 *New Zealand Line* has a short train at Wareham station in the early 1960s. Photograph by Bill Reed.

Above WESTON-SUPER-MARE STATION – NO. 6109
A train is led away from Weston-super-Mare station by Collett 6100 Class no. 6109 on 12th September 1952. A short branch from the Bristol & Exeter Railway originally served the town from 1841. This was horse-drawn at first and steam was not used until the early 1850s. Improved facilities were built in the mid-1860s and the branch became a through route in 1884. No. 6109 was Bristol-allocated when pictured. Photograph by B.W.L. Brooksbank.

Opposite WESTBURY STATION – NO. 4924
The Wilts, Somerset & Weymouth Railway reached Westbury which was initially used as the location of a terminus on the first section from Chippenham. This opened in early September 1848 yet two years elapsed before the next short section to Frome was ready. The line to Weymouth did not reach completion until 1857, whilst a year earlier Westbury became a junction for a branch to Salisbury. At the turn of the century, Westbury saw a short line added to the route from Reading which was intended to shorten the journey between London and Taunton. Westbury station was rebuilt to serve all these lines as a result. Collett 'Hall' no. 4924 *Eydon Hall* has an express from one at Westbury c. 1960. The engine is in particularly good condition, suggesting an ex-works service originating at Swindon. Photograph by Bill Reed.

Above WEYMOUTH – NO. 7917
Hawksworth 'Modified Hall' Class no. 7917 *North Aston Hall* was amongst the last of a batch of 29 engines built to the design at Swindon Works between 1948 and 1950. A further ten appeared before construction ceased. In the mid-1950s, no. 7917 has a local train at Weymouth. The engine was Westbury-allocated throughout the decade. Photograph by Bill Reed.

Opposite WEYMOUTH HARBOUR – NO. 1368
Weymouth and the surrounding area provided a harbour for boats following the Norman Conquest. Early trade focused on wine and wool, as well as soldiers moving between England and France. Weymouth continued to be important through the centuries and this was bolstered by the arrival of the Wilts, Somerset & Weymouth Railway in 1857. Operated at this time by the GWR, the company built the Weymouth Harbour Tramway in 1865 to serve the docks, running through the streets on a branch leaving the original route north of the station. Weymouth Quay station was opened in 1889 to serve cross-channel boat services. The line called for specialised locomotives and Collett's version was the 1366 Class built in 1934. No. 1368 was the third of six built at Swindon and a trio was based at Weymouth, with the engine a long-term resident. Seen at the harbour on 1st September 1959, no. 1368 was employed to May 1962 but worked for another two years at Wadebridge after replacing the Beattie 0298 Class. Much of the rail system has been recently removed from Weymouth docks though regular services had ceased in the late 1980s. Photograph by Geoff Warnes.

Below WEYMOUTH HARBOUR – NO. 1367
Collett 1366 Class no. 1367 – unofficially named *Percy's Pet* – is at Weymouth Harbour during the mid-1950s. The engine had a brief spell at Swindon in 1953 and also worked the Wadebridge line in the early 1960s. Photograph by Bill Reed.

Above **WEYMOUTH SHED – NO. 30865**
At Weymouth shed for servicing in the late 1950s is Maunsell 'Lord Nelson' Class no. 30865 *Sir John Hawkins*. The L&SWR was granted running powers to Weymouth via a connection at Dorchester on the line from Bournemouth where no. 30865 was based. Photograph by Bill Reed.

Opposite above **WEYMOUTH – NO. 6994**
'Modified Hall' no. 6994 *Baggrave Hall* is north of Weymouth station in the mid-1950s. Photograph by Bill Reed.

Below **WEYMOUTH – NO. 1368**
A train for Weymouth Quay is on Commercial Road with no. 1368 in September 1959. Photograph from the Dave Cobbe Collection courtesy Rail Photoprints.

Opposite above WHITSTONE & BRIDGERULE STATION – NO. 31859
The final section of the L&SWR's route to Bude was completed in 1898 and ran from Holsworthy via Whitstone and Bridgerule. The station for the last two places was situated between them and was the location where a passing point was established, the remainder of the route being single track. Maunsell N Class 2-6-0 no. 31859 has reached Whitstone & Bridgerule with the 13.15 Bude to Exmouth Junction service on 9th July 1964. The engine was based at the last-mentioned place and was condemned there in September. Whitstone & Bridgerule station was closed with the line in 1966 and passed into private hands. Interestingly, the platform canopy appears to still be present. Photograph by Dave Cobbe courtesy Rail Photoprints.

Opposite below WILTON SOUTH STATION – NO. 30826
A short distance west of Salisbury, Wilton South station was built on the Salisbury & Yeovil Railway and opened in early May 1859. A GWR station was present nearby and had been in use for three years at that time. Following Nationalisation, the pair were distinguished with the suffixes South and North respectively. Wilton South was also the location where the engines used on the 'Devon Belle' changed and avoided congestion at Salisbury. S15 Class 4-6-0 no. 30826 has the 12.55 Templecombe to Salisbury train on the main line, whilst 'Merchant Navy' Pacific no. 35007 *Aberdeen Commonwealth* waits off the left for the 'Devon Belle' in order to take over. Photograph courtesy Rail Photoprints.

Below WHATLEY QUARRY
West of Frome, Whatley quarry was established to exploit limestone deposits. Andrew Barclay 0-4-0ST, works no. 969, *Medway* is in charge of a hopper train receiving the mineral on 30th December 1952. The quarry is still operational and retains a rail connection to part of the Frome-Radstock branch. Photograph by Hugh Ballantyne courtesy Rail Photoprints.

Above YELVERTON STATION – NO. 4410
Churchward introduced the 4400 Class 2-6-2T to cater for branch line traffic, particularly those with adverse gradients. The type had relatively small driving wheels but had standard parts in common with the contemporary 4500 Class. Ten of the 4400s were built at Wolverhampton and a solitary example appeared from Swindon. No. 4410 was the final engine of the class and in service during June 1906. Reaching the end of a nearly 50-year career here, no. 4410 is at Yelverton station with a Princeton branch train around 1953. Yelverton was on the line from Plymouth to Tavistock which was laid by the South Devon & Tavistock Railway in 1859. A late addition to the route, Yelverton opened in 1885 after the GWR completed a branch to Princetown, 10 miles to the north east, two years earlier. Photograph courtesy Rail-Online.

Opposite above YELVERTON STATION – NO. 4568
Churchward 4500 Class no. 4568 has a branch service at Yelverton station on 21st February 1956. The locomotive was constructed at Swindon Works in October 1924 and had a career spanning 34 years. At the time of the image, no. 4568 was allocated to Plymouth but in June 1956 moved on to Newton Abbot. The final move took the engine to Penzance in September 1958. Photograph by G.H. Hunt courtesy Colour-Rail.

Opposite below YEOVIL JUNCTION STATION – NO. 30131
Mid-1860 was a busy period for the L&SWR as the line from Salisbury reached completion in June and in July the extension to Exeter began operations. Both lines passed Yeovil to the south and a connection was made to the town where Hendford station was shared with the Bristol & Exeter Railway, which had a branch from their main line. Yeovil Junction station opened in 1860 to serve the connection and from 1861 diverted trains to Yeovil Town. Drummond M7 Class no. 30131 has a branch train at Yeovil Junction during August 1961. Photograph by Bill Reed.

Above **YEOVIL TOWN STATION – NO. 5416**
An 0-6-0PT locomotive capable of moderate speeds on local passenger trains was necessary in the early 1930s. The design for William Dean's 2021 Class was updated and given 5 ft 2 in. diameter driving wheels. One class member was rebuilt to 5400 Class specifications before the production run started in 1930. This lasted to 1932 when 20 were in traffic and in 1935 the final five followed. No. 5416 was erected at Swindon Works in June 1932. The engine is at Yeovil Town station on 21st April 1963 when just four months away from being sent to the scrapyard. No. 5416 has the Yeovil Town to Yeovil Pen Mill shuttle train. The latter was on the Wilts, Somerset & Weymouth line and connected with the Yeovil branch of the Bristol & Exeter Railway. Photograph by Hugh Ballantyne courtesy Rail Photoprints.

Below **YEOVIL JUNCTION STATION – NO. 34061**
The 12.36 Salisbury to Exeter at Yeovil Junction station on 26th March 1964. Bulleid 'Battle of Britain' Pacific no. 34061 *73 Squadron* is at the head of the service. Built at Brighton Works in April 1947, the locomotive survived to August 1964. Spending much of the 1950s at Exmouth Junction, no. 34061 was in the South East during the early 1960s, but returned to Exmouth from November 1963 to June 1964. The engine's last two months passed at Eastleigh. Photograph by R. Patterson courtesy Colour-Rail.

Above YEOVIL TOWN STATION – NO. 30125
At Yeovil Town station with a local service on 6th June 1961 is Drummond M7 Class no. 30125. Just over six months earlier, the locomotive had been allocated to Exmouth Junction shed and has the '72A' code here. No. 30125 was in service there until condemned in late December 1962. In mid-1963 the depot was transferred to the Western Region, becoming '83D' and continued to service steam locomotives until 1967. Photograph by Tony Cooke courtesy Colour-Rail.

Opposite YEOVIL TOWN STATION – NO. 5504
As mentioned, Yeovil Hendford station was briefly the main station at Yeovil before the opening of Yeovil Town in 1861. This was used for 106 years before closure and the site has been cleared for redevelopment. Along with branch services, the station ran shuttles between both Pen Mill and Yeovil Junction. Collett 4575 Class no. 5504 has a local train at Yeovil Town on 31st August 1959. Built in May 1927, the locomotive was condemned in October 1960. Photograph by Geoff Warnes.

BIBLIOGRAPHY

Allen, Cecil J. *Titled Trains of Great Britain.* 1983.
Bradley, D.L. *Locomotives of the L.S.W.R.: Part 2.* 1967.
Bradley, D.L. *Locomotives of the Southern Railway: Part 1.* 1975.
Bradley, D.L. *Locomotives of the Southern Railway: Part 2.* 1976.
British Rail Main Line Gradient Profiles.
Griffiths, Roger and Paul Smith. *The Directory of British Engine Sheds and Principal Locomotive Servicing Points: 1 Southern England, the Midlands, East Anglia and Wales.* 1999.
Quick, Michael. *Railway Passenger Stations in Great Britain: A Chronology.* 2009.
RCTS. *A Detailed History of British Railways Standard Steam Locomotives: Volume One Background to Standardisation and the Pacific Classes.* 2007.
RCTS. *A Detailed History of British Railways Standard Steam Locomotives: Volume Two: The 4-6-0 and 2-6-0 Classes.* 2003.
RCTS. *A Detailed History of British Railways Standard Steam Locomotives: Volume Three: The Tank Engine Classes.* 2007.
RCTS. *A Detailed History of British Railways Standard Steam Locomotives: Volume Four The 9F 2-10-0 Class.* 2008.
RCTS. *Locomotives of the Great Western Railway: Parts One to Twelve.* 1951-1974.
St John Thomas, David. *A Regional History of the Railways of Great Britain: Volume 1 – The West Country.* 1973.
Walmsley, Tony. *Shed by Shed: Part Six – Western.* 2009.
White, H.P. *A Regional History of the Railways of Great Britain: Volume 2 – Southern England.* 1983.

Also available from Great Northern

The Last Years of Yorkshire Steam
The Golden Age of Yorkshire Railways
Gresley's A3s
Peppercorn's Pacifics
London Midland Steam 1948-1966
The Last Years of North East Steam
British Railways Standard Pacifics
Western Steam 1948-1966
The Last Years of North West Steam
Gresley's V2s
Southern Steam 1948-1967
Yorkshire Steam 1948-1967
Gresley's A4s

Gresley's B17s
The Last Years of West Midlands Steam
East Midlands Steam 1950-1966
Thompson's B1s
The Glorious Years of the LNER
Scottish Steam 1948-1967
North East Steam 1948-1968
The Last Years of London Steam
Gresley's D49s
The Glorious Years of the GNR
East of England Steam 1948-1963
Stanier's Jubilees
The Glorious Years of the LMS

visit www.*greatnorthernbooks.co.uk* for details.